RAND

Defense Working Capital Fund Pricing Policies

Insights from the Defense Finance and Accounting Service

Edward G. Keating, Susan M. Gates

Prepared for the
Defense Finance and Accounting Service

National Defense Research Institute

Preface

The Defense Finance and Accounting Service (DFAS) is a defense agency created in 1991 to provide finance and accounting services (e.g., personnel and contract payments, trial balances, travel voucher handling) for the Department of Defense (DoD). DFAS agglomerated what had been service-specific facilities and personnel providing such services. DFAS is also a Defense Working Capital Fund (DWCF) activity, which recovers its operating costs through customer fees.

The DFAS asked RAND to examine historical trends and patterns in DFAS's performance. The goal is to help DFAS leadership identify possible approaches to improve performance. This report integrates and summarizes RAND's Fiscal Year (FY) 1998 research for DFAS and presents pricing policy implications that may be applicable to other DoD and governmental entities.

This research was conducted in the Forces and Resources Policy Center of RAND's National Defense Research Institute, a federally funded research and development center sponsored by the Office of the Secretary of Defense, the Joint Staff, the unified commands, and the defense agencies. This report will be of interest to DoD policymakers and researchers interested in infrastructure and intra-DoD pricing policies.

Contents

Figures

Tables

Summary

Background and Purpose

The Defense Finance and Accounting Service (DFAS) was created in 1991 to consolidate military sevice-specific accounting and finance operations. The goal of this consolidation was to reduce the number of disparate finance and accounting related systems, and thereby reduce Department of Defense expenditures on such services.

This report analyzes DFAS's performance and recommends approaches for improving it. We used DFAS's Resource Analysis Decision Support System (RADSS) data, which provides monthly information on DFAS expenditures and workloads, to analyze performance.

DFAS Revenues and Costs

The DFAS is a Defense Working Capital Fund (DWCF) entity. Under DWCF policy, DFAS prices are typically set two years in advance. Prices are set equal to average cost at expected demand levels. The DWCF-dictated price equals the expected total cost of producing the expected number of work units divided by the expected number of work units. With this linear pricing structure, DFAS revenue increases and decreases in exact proportion to actual DFAS workload.

In contrast to revenues, DFAS's costs do not appear to move in close correspondence to the number of work units. The expected total costs used to set prices appear to include both fixed costs (those incurred regardless of the number of work units produced) and incremental costs (those that change as output levels change). Some costs may be "fixed" because of:

- observable, technology-related reasons
- the costs of creating and maintaining finance and accounting systems driving the DFAS cost structure
- a reluctance or inability to easily reduce civilian employee costs as workload falls.

Implications

The nonlinear nature of costs is in direct contrast with DWCF pricing policies. DWCF-dictated prices, based upon projected fixed and incremental costs, appear far higher than the incremental costs of producing work units. Such prices may not provide appropriate incentive for customers to use DFAS services, thus leading to inadequate revenue. Prices must then be raised to compensate for revenue shortfalls, and costs must be reduced. Higher prices can further impact workload levels. Cost reductions have not kept pace with revenue decreases, and it may be difficult for them ever to do so.

We believe DFAS's current pricing structure is unsustainable. In particular, as long as demand for DFAS's services continues to fall faster than planned, the current DWCF pricing rules imply DFAS's revenues will chronically fall short of its costs. DFAS's pricing structure needs to be changed. The nonlinear tactics employed by private-sector firms facing such a cost structure may prove to be useful models for DFAS.

Recommendations

Our recommendation is that DFAS experiment with a simple nonlinear pricing structure. Under such an arrangement, DFAS would receive "open the door" transfer payments from its customers that reflect its fixed costs. DFAS would also assess incremental fees per unit of workload that would be lower than current DFAS prices. Such a structure would give its customers more appropriate incentives with respect to how much workload to give DFAS. Both transfer and incremental fees could vary among customers, as well as outputs.

We do not wish to understate the challenges associated with changing DFAS's pricing algorithm. Dividing DFAS's fixed costs effectively among its customers will require careful analysis. While acknowledging some costs to be invariant to output, or fixed, DFAS should continue efforts to reduce these costs. Similarly, prudent analysis will be needed to determine appropriate incremental fees.

Acknowledgments

This research was sponsored by Bruce M. Carnes, DFAS's Director for Resource Management. The authors especially thank Carol Robertson of DFAS for assistance with, and helpful comments on, this research. The authors also thank DFAS's Wanda Brandon, Joel Carey, Scott Chellberg, Michael Dugan, Jim Dundon, Christy Edwards, David Harris, Jim Howard, Steve Krawczel, Bill List, Ed Malone, Keith Mooney, Roxane Nowling, Melanie Reed, David Sundby, and Greg Trygg. We also thank the large number of DFAS regional center personnel we have met.

Susan Hosek of RAND supervised this research. Marygail Brauner and Jim Dertouzos provided helpful reviews of this work. Rodger Madison provided extensive computer assistance. Gordon Lee served as a Communications Analyst on this project. Jennifer Pace provided research assistance. Ron Key, Chris Myrick, and Miriam Polon edited this document. The authors also thank RAND colleagues Laura Baldwin, Joe Bolten, Dick Buddin, Frank Camm, David Chu, Marc Elliott, David Gompert, Gene Gritton, Michele Guemes, Chris Hanks, Jan Hanley, Jeanne Heller, Lisa Hochman, Michael Kennedy, Rebecca Kilburn, Janet Kleinman, Nancy Moore, Ellen Pint, Peter Reuter, Irene Sanchez, Regina Sandberg, Mark Wang, Stephanie Williamson, and Benson Wong for their help.

This research was briefed to Gary Amlin, the acting director of DFAS, on August 31, 1998. This research was presented at a RAND Logistics lunch on October 2, 1998. Also, it was briefed at the DFAS Resource Management Conference in Arlington, Virginia on November 19, 1998. An earlier version of this research was presented to a RAND Forces and Resources Policy seminar on May 7, 1998. Comments by seminar participants were most appreciated.

Of course, remaining errors are the authors' responsibility.

Abbreviations

ABC	Activity-Based Costing
ADP	Automatic Data Processing
AMC	Army Materiel Command
CORM	Commission on Roles and Missions
DAO	Defense Accounting Office
DBOF	Defense Business Operations Fund
DCPS	Defense Civilian Pay System
DF	Degrees of Freedom
DeCA	Defense Commissary Agency
DFAS	Defense Finance and Accounting Service
DISA	Defense Information Systems Agency
DLA	Defense Logistics Agency
DMDC	Defense Manpower Data Center
DoD	Department of Defense
DWCF	Defense Working Capital Fund
FMS	Foreign Military Sales
FSO	Financial Systems Organization
FY	Fiscal Year
GAO	General Accounting Office
GS	General Schedule
MEO	Most Efficient Organization
MOCAS	Mechanization of Contract Administration Services
NPR	National Performance Review
OMB	Office of Management and Budget
OOS	Out of Service
OPLOC	Operating Location
PMIS	Performance Management Information System
QDR	Quadrennial Defense Review
RADSS	Resource Analysis Decision Support System
RIF	Reduction in Force
SAMMS	Standard Automated Material Management System
SE	Standard Error
SS	Sum of Squares
U.S.	United States
Y2K	Year 2000

1. Introduction

Background

The Defense Finance and Accounting Service (DFAS) provides a variety of services to Department of Defense (DoD) customers, such as payroll, bill payment, and generation of accounting statements. DFAS was created in 1991 through the consolidation of military service-specific accounting and finance operations.[1] The goal was to reduce DoD expenditures on, and improve the quality of, such services. Prior to this consolidation, each military service had a relatively autonomous finance and accounting center, and substantial support activities were conducted at the installation level. The finance and accounting headquarters for the Air Force, Army, Marines, and Navy were in Denver, Colorado; Indianapolis, Indiana; Kansas City, Missouri; and Cleveland, Ohio, respectively. The finance and accounting activities for DoD agencies were centered in Columbus, Ohio. The services also had a multitude of highly disparate accounting and finance systems in operation.

The DFAS has undertaken an effort to increase standardization in accounting and finance-related data systems across the DoD. Since the creation of DFAS in 1991, there have been reductions in the number of finance and accounting sites and personnel. It is estimated that the DoD had 324 separate finance and accounting systems before DFAS; as of Fiscal Year (FY) 1998, the estimated total was down to 121.

DFAS currently has its headquarters in Arlington, Virginia. There are five regional centers and 18 operating locations (OPLOCs) in the United States. Each OPLOC is affiliated with a particular regional center. Table 1.1 provides DFAS's regional center/OPLOC arrangements as of July 1, 1998. We refer to the combination of a regional center and its associated OPLOCs as a "region." Though the DFAS regions now do some work for other services, the pre-DFAS work patterns have persisted to a large extent.

---

[1]General Accounting Office (GAO) report NSIAD/AIMD-97-61 discusses the development of DFAS.

Table 1.1

DFAS Regional Centers and OPLOCs

Regional Center	Associated OPLOCs	Primary Customer
Cleveland, OH	Charleston, SC Honolulu, HI[a] Norfolk, VA Oakland, CA Pensacola, FL San Diego, CA	Navy
Columbus, OH	None	DoD Agencies
Denver, CO	Dayton, OH Limestone, ME Omaha, NE San Antonio, TX San Bernardino, CA	Air Force
Indianapolis, IN	Lawton, OK Lexington, KY Orlando, FL Rock Island, IL Rome, NY Seaside, CA St. Louis, MO	Army
Kansas City, MO	None	Marines

Source: DFAS Web site—http://www.dfas.mil.

[a]DFAS also has a satellite facility in Japan that reports through the Honolulu OPLOC.

Objective

The DFAS asked RAND to examine historical trends and patterns in DFAS's performance. The objective is to identify possible approaches for improving DFAS performance.

Approach

This study is based on DFAS's Resource Analysis Decision Support System (RADSS) data, which calibrate DFAS costs and workloads over time. We used these data to evaluate DFAS costs and prices relative to workload. We developed a standardized index to compare actual costs to expected costs. We also performed regressions to verify our findings about the cost/workload relationship.

Organization of the Report

The remainder of this report is organized into four additional sections.

Section 2 discusses DFAS pricing and revenues and DFAS costs. Section 3 reviews the analysis of DFAS performance using cost/workload indices. It also describes the DFAS cost structure and how it relates to DFAS performance. Section 4 examines the implications of maintaining the current pricing structure, and discusses possible approaches to improving DFAS performance. Section 5 provides our conclusions about DFAS performance and our recommendations on how to improve it.

2. DFAS Revenues and Costs

Prices and Revenues

The DFAS is a Defense Working Capital Fund (DWCF) entity. The DoD has a number of DWCF entities, such as the Defense Logistics Agency (DLA) and the Defense Information Systems Agency (DISA). The military services also have their own DWCF-type arrangements, such as the Army Stock Fund. DFAS is just one example of a DWCF-funded organization in the DoD.

Under DWCF policies, DFAS (like other working capital fund entities) receives payments from its customers based on the number of work units it processes. Each of these work units has a price assigned to it. A price per travel voucher, for instance, is set ahead of time; DFAS's revenue from travel vouchers is that price multiplied by the number of travel voucher work units DFAS processes.

The DFAS prices, as dictated by DWCF policy, are set two years in advance. They are designed to result in revenues equal to costs at expected demand levels.[1] In order to accomplish this goal, prices are set equal to average cost at expected demand levels; the DWCF-dictated price equals the expected total cost of producing the expected number of work units divided by that number of work units.

Once these prices are established, DFAS's actual revenue is a linear function of the number of work units purchased by customers. One additional work unit increases DFAS's revenue by the prespecified price of that type of work unit. One less work unit decreases DFAS's revenue by the prespecified price of that type of work unit.

Costs

Unlike many governmental organizations, DFAS has a useful data set that calibrates its costs and workload over time. The RADSS data provide monthly information on costs and work units (e.g., the number of commercial invoices processed) for each designated financial service and location. DFAS has defined

---

[1]GAO/AIMD-97-134 describes the DWCF price-setting process.

the services it provides as a set of outputs, as listed in Table 2.1. Almost all DFAS regional costs are allocated to some output. We cannot, however, verify the validity of the allocation that has occurred. One concern might be, for example, that overhead is allocated inappropriately within DFAS.

Overall, we believe the RADSS data set is a valuable tool to provide insights on DFAS costs and performance. It would be considerably harder to evaluate the performance of, and draw conclusions from, an agency that lacked such rich, detailed data. That said, there were also some problems in RADSS with missing and erroneous data. We made our best efforts at addressing these problems through discussions with the providers of the data, as well as with regional personnel.

Our analysis was performed on RADSS monthly cost and workload data for each of DFAS's five regions dating back to mid-1995, except as noted otherwise.[2]

Figure 2.1 breaks up DFAS inflation-adjusted expenditures by region. Indianapolis, with 34 percent of expenditures, is the largest DFAS region; Kansas City, with only 4 percent of expenditures, is considerably smaller than the other

Table 2.1

DFAS Outputs

DFAS Outputs
Civilian Pay
Commercial Invoices
Contract Invoices (MOCAS)
Contract Invoices (SAMMS)
Direct Billable Hours
Finance & Accounting Commissary
Foreign Military Sales
Military Active Pay Accounts
Military Pay Incremental
Military Reserve Pay Accounts
Military Retired Pay Accounts
Monthly Trial Balances
Out-of-Service Debt Cases
Support to Others
Transportation Bills
Travel Vouchers

[2]We have not analyzed DFAS headquarters costs.

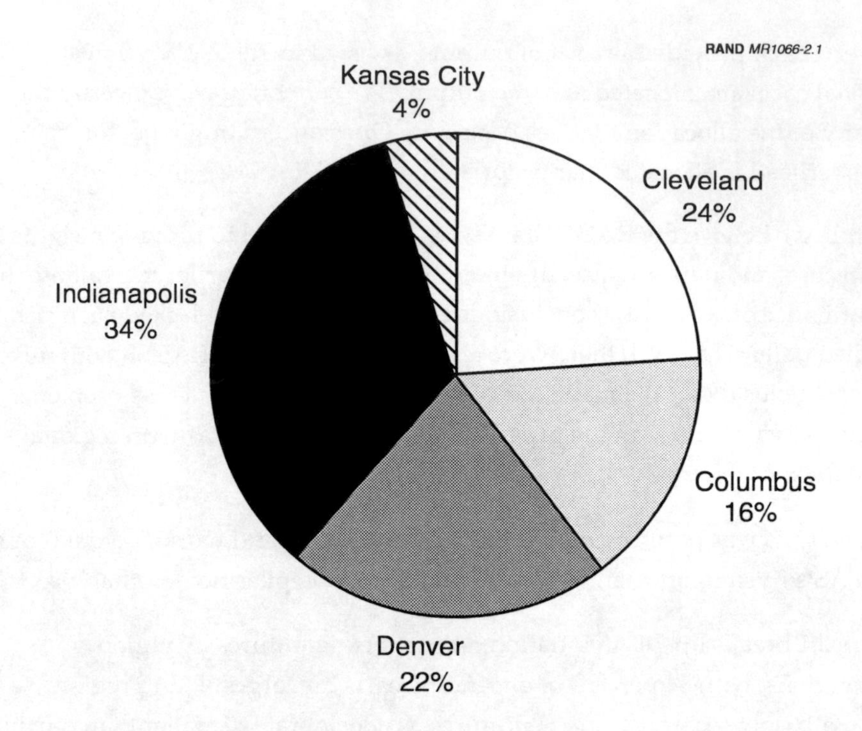

Figure 2.1—DFAS Expenditures by Region

regions.[3] There have been no marked changes in the regions' relative expenditure shares over the last few years.

Figure 2.2 breaks up DFAS expenditures by output for all regions. While DFAS produces a number of outputs, monthly trial balances and commercial invoices combine to represent more than half of DFAS's regional costs.

Figure 2.3 shows the total monthly expenditures in all DFAS regions in inflation-adjusted FY96 dollars. DFAS regions, in total, have typically expended about $135 million per month, or an annualized total of about $1.6 billion. This graph shows there has been some tendency toward declining real expenditures. For example, the average monthly expenditure in 1996 was about $139 million, while the average monthly expenditure during the first five months of 1998 was about $122 million in FY96 dollars.

In contrast to revenues, DFAS's costs do not appear to move in close correspondence to the number of work units. This is illustrated in Figure 2.4, which shows DFAS Region A civilian pay account costs and work units. The

---

[3]Figures 2.1 and 2.2 cover RADSS expenditure data, corrected for inflation, between August 1995 and May 1998, but excluding October 1997 (due to the omission of that month in the Denver region expenditure data we received).

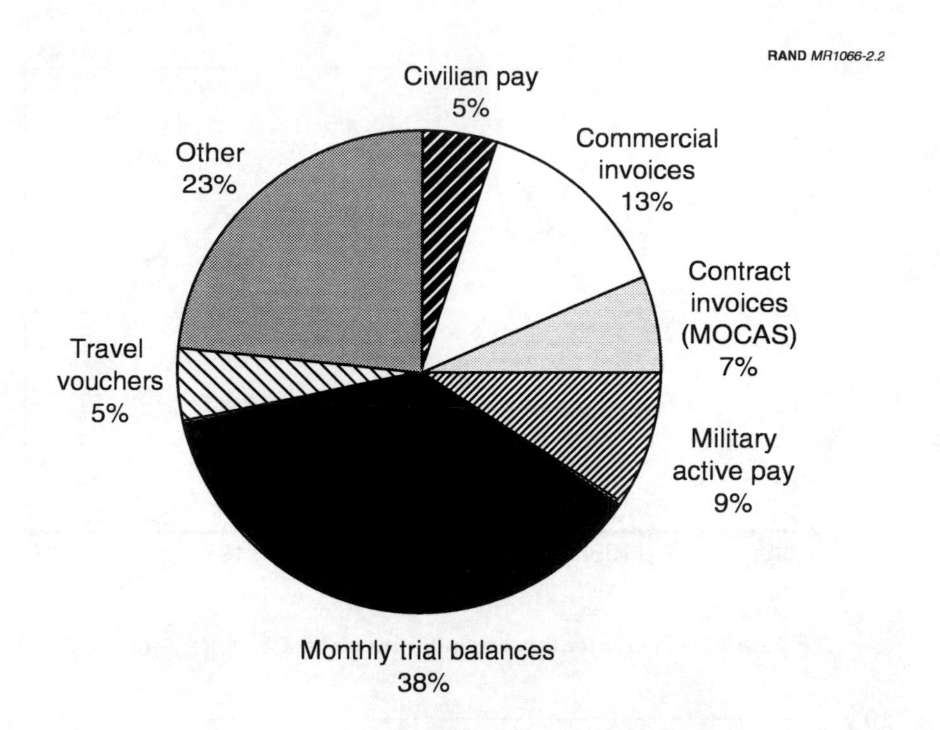

Figure 2.2—DFAS Regions' Expenditures by Output

solid line shows inflation-corrected Region A civilian pay account expenditures. The broken line shows Region A's civilian pay account workload. The noteworthy characteristic of Figure 2.4 is that, in early FY 1997, Region A's civilian pay account workload surged with little, if any, impact on its civilian pay account costs.

Figure 2.5 also shows that DFAS's costs do not appear to move in close correspondence with the number of work units. The solid line is inflation-corrected Region B's military active pay account expenditures. The broken line is Region B's military active pay account workload. Region B's military active pay account workload has fallen while costs have shown no apparent trend. DFAS personnel told us that Region B's primary customer changed military pay account systems during the first six months of FY 1998 and that this transition retarded cost savings.

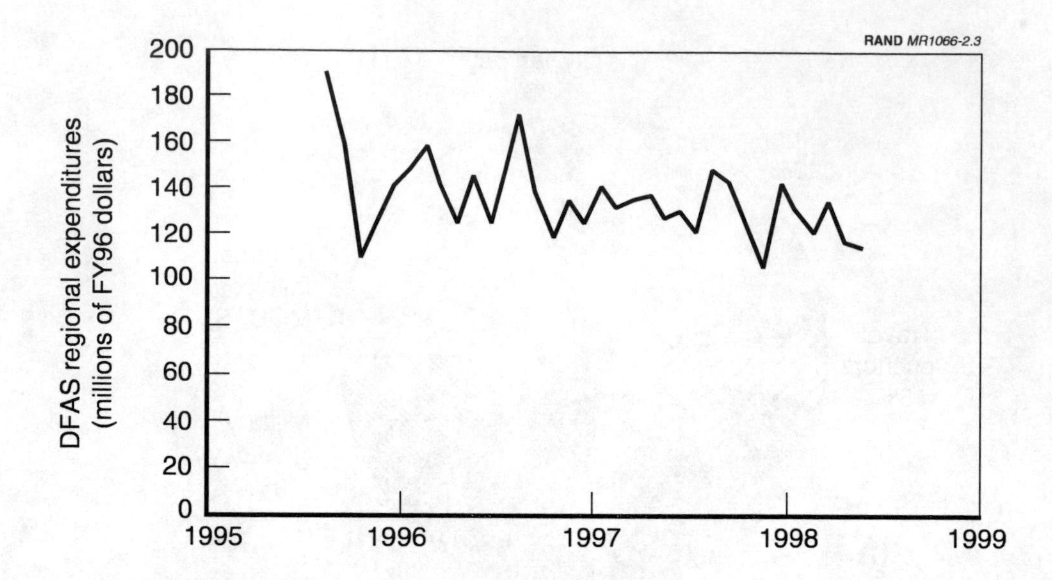

Figure 2.3—Total Monthly Expenditures of All DFAS Regions

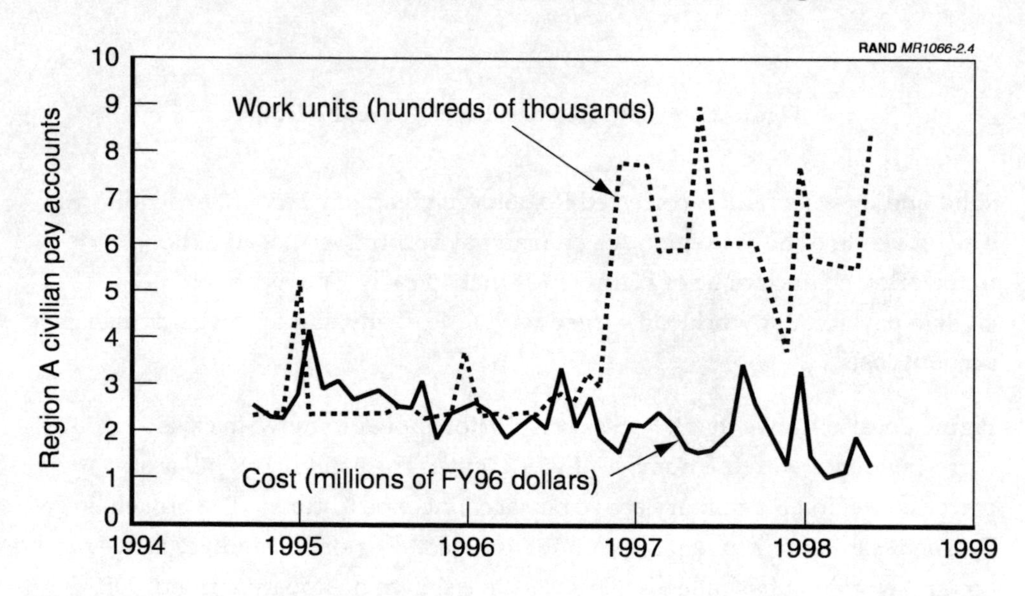

Figure 2.4—Region A Civilian Pay Costs and Work Units

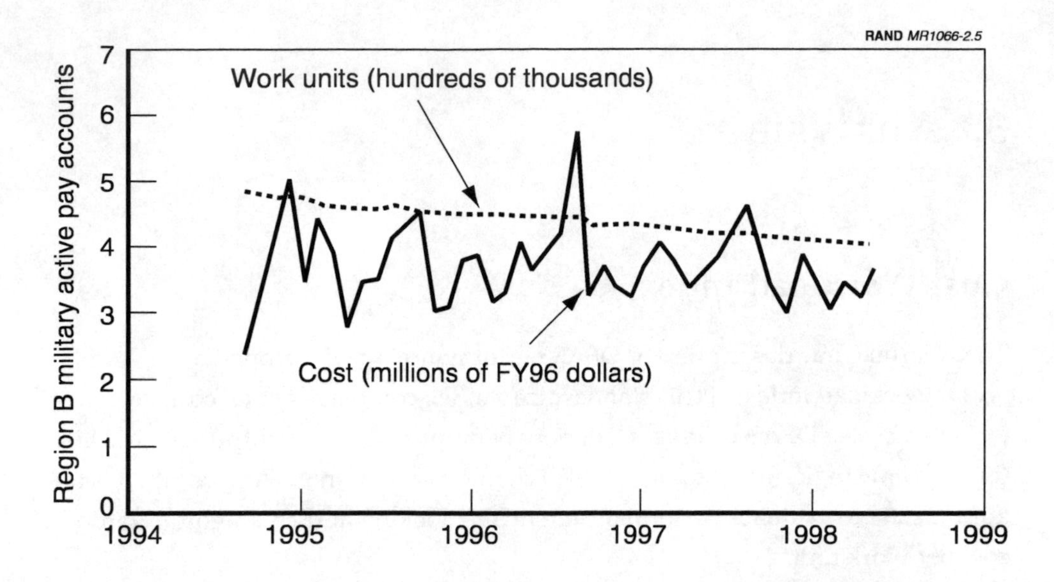

Figure 2.5—Region B Military Active Pay Costs and Work Units

3. Analysis

Cost/Workload Index

To undertake our description of DFAS performance, we developed a "cost/workload index." This standardized index compares actual costs to expected costs. DFAS produces a variety of outputs, and the output mix of DFAS (or a specific region, center, or OPLOC) changes each month. As a result, it is not appropriate to evaluate performance trends by looking at the straightforward cost per work unit.

The cost/workload index we have created accounts for output mix variation by calculating the weighted average of the expected costs of each output. If actual costs equal expected costs in a particular time period, the index has a value of 1. When actual costs are less than expected costs, the index has a value less than 1.

The expected cost for a particular output is the average cost of producing that output over the data's full time period for a specific entity (DFAS, a region, a center, or an OPLOC). As a result, the average value of the cost/workload index over the data's full time period for a specific entity will always be equal to 1. In a specific month, however, the index value may be above or below 1.

Index values can be computed at different levels of aggregation, e.g., all DFAS regions, specific regions, or specific outputs in specific regions. If one focuses on a specific output in a specific region, the cost/workload index will be equivalent to expenditures per work unit.

Table 3.1 works through a simple, purely illustrative example of how a cost/workload index is calculated.

We begin by calculating average costs per work unit over the whole period covered by the data. Such average costs are then multiplied by actual work units to form expected costs for each month. The cost/workload index for a month, then, is the ratio of actual costs to expected costs, with an intertemporally declining ratio suggesting performance improvement. In Table 3.1's example, measured performance worsens slightly from November 1998 to December 1998 (accounting for the change in workload mix between the two months).

Table 3.1

An Example of Cost/Workload Index Calculation

Month	Washing Work Units	Total Washing Costs	Drying Work Units	Total Drying Costs
November 1998	100	$1,000,000	50	$500,000
December 1998	75	$900,000	75	$650,000
Average Cost per Work Unit in Data:		$10,857		$9,200

Month	Actual Washing Costs	Expected Washing Costs	Actual Drying Costs	Expected Drying Costs
November 1998	$1,000,000	$1,085,714	$500,000	$460,000
December 1998	$900,000	$814,286	$650,000	$690,000

Month	Aggregate Cost/Workload Index
November 1998	0.97
December 1998	1.03

We chose to develop this cost/workload index procedure instead of using the simpler notion of aggregate average cost per work unit. The problem with measuring aggregate average cost per work unit is that workload composition changes (e.g., adding proportionally more high-cost outputs) can artificially and misleadingly change the aggregate average cost per work unit. It is a proverbial apples-to-oranges comparison to simply add work units across different outputs.

We want to be careful to convey the limitations of cost/workload indices. First, these indices say nothing about the quality of work undertaken. Work units are simply tallied; provision quality, such as accuracy and timeliness, is not considered. More generally, RADSS data do not consider performance quality.

Second, there are challenges in using cost/workload indices to compare the efficiency levels of different DFAS regions. As noted above, DFAS regions have traditionally been affiliated with specific branches of the armed services, and the different services have different approaches and procedures for handling similar outputs, such as travel vouchers. We have therefore created our cost/workload indices only within specific regions. Our indices might be used to evaluate which regions are improving and which are not, thereby revealing gross differences in performance trends. They do not, however, indicate whether the absolute level of performance in one region is better than in another. This is because, by construction, each region's average index value is one and only intra-regional average expenditures per work unit are utilized.

Third, the way DFAS regional overhead is allocated exaggerates the costs of DFAS's regional centers versus its OPLOCs. The costs of regional center employees who assist the entire region, such as the regional director of resource management, are assigned only to the center and attributed only to the center's work units. This overhead allocation problem implies that the level of efficiency of centers versus OPLOCs cannot be meaningfully assessed.

Finally, DFAS cost/workload indices shed no light on the comparative efficiency of DFAS versus private sector providers.

Figure 3.1 uses our cost/workload indexing procedure to summarize DFAS regions' cost/workload performances from August 1995 to May 1998 (excluding October 1997). Though the data are choppy, the overall trend seems to be moderately favorable, i.e., downward sloping.[1] The horizontal line at 1.0 reflects the fact that the average index value is 1, by construction.

We developed this aggregate cost/workload index by computing each region's average cost per work unit over the entire period of our data for each type of output. Next, we used these regional output averages to compute expected costs for each region for each month. The displayed index is the ratio of regions' actual costs to their expected costs. By construction, the average ratio is 1, but

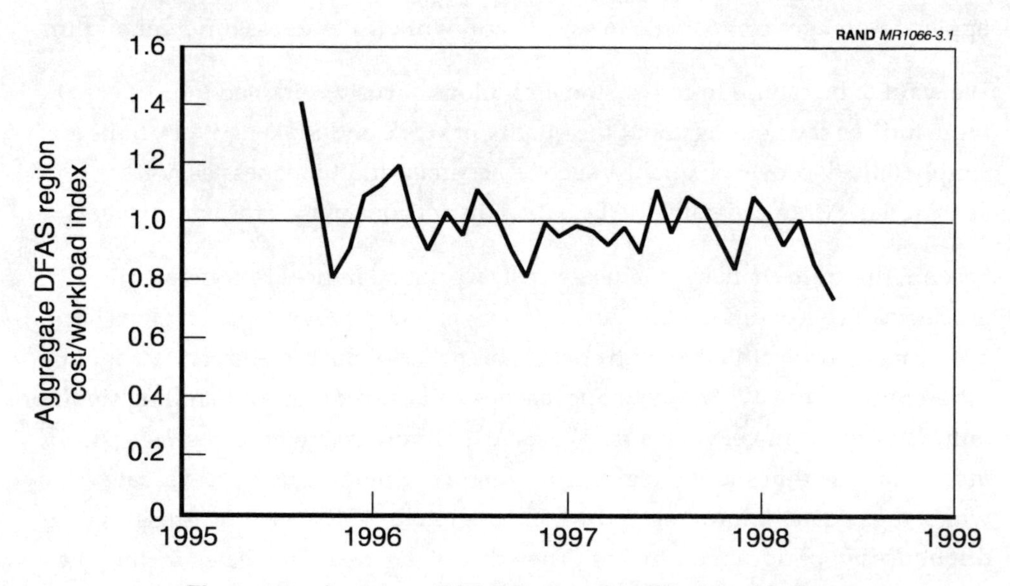

Figure 3.1—Aggregate DFAS Region Cost/Workload Index

---

[1]Appendix A provides regression analysis complementing the assertions made in this chapter. For example, we confirm that Figure 3.1's data have a downward trend if one assumes each month's observation is a statistically independent observation.

this type of display shows performance time trends with downward slopes indicating improvement.

Costs incurred by DFAS headquarters are not included in Figure 3.1. More generally, we could not include output categories (e.g., "Support to Others" or "Unassigned") where work units are not tabulated. A display of this sort might be misleading if important changes are occurring in untabulated areas. Fortunately, such categories do not represent a large percentage of DFAS costs.

Figure 3.1's display aggregates the performance of the five DFAS regions. It is possible to separate the data and calculate separate indices for each region. These calculations reveal differences in performance trends. In Figure 3.2 we show performance trends for two regions that we label C and D.[2] Though the index values vary considerably month-to-month, Region C's performance seems to have improved, as shown by the general tendency of the Region C index values to decline. Note, for example, that eight of the last nine Region C months have index values below 1. Results, until recently, have been somewhat less positive for Region D. Region D had worsening, though less variable, performance until the start of 1998; it only roughly reachieved its level of performance at the start of FY 1995 (the index value has dropped back toward the FY 1995 level) by May 1998.

Regional data of this sort include the regional headquarters plus all affiliated operating locations and defense accounting offices (DAOs).

Within regions, there is variability in performance across outputs. Each region produces 7–9 different finance and accounting outputs. The top graph in Figure 3.3, for instance, shows improvement in Region E's military reserve pay account performance, with the last seven months below 1. (The negative value in July 1995 reflects a negative expenditure value in RADSS for that month; this data anomaly is excluded from Appendix A's estimations.) On the other hand, Region E's transportation bill performance appears to have worsened, with the last six months above 1.

As shown in this section, the cost/workload index procedure can be run on a number of different levels of aggregation in the data. Different types of management questions require different levels of analysis of the data.

---

[2]In this report, we do not specifically identify the regions. Instead, we give them generic labels. This masking does not obfuscate our point that there is performance trend heterogeneity across regions.

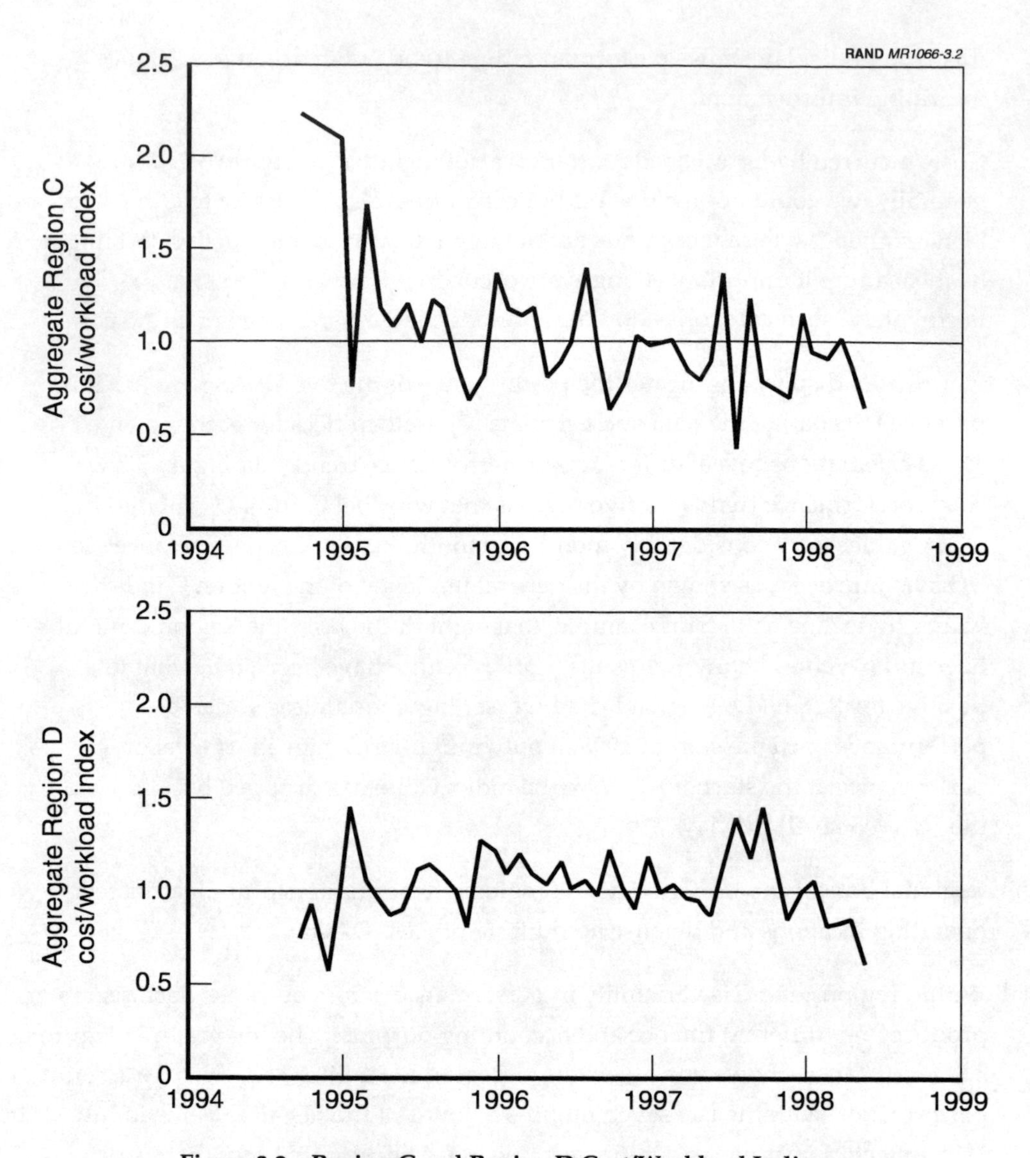

Figure 3.2—Region C and Region D Cost/Workload Indices

Fixed Costs

Why do costs not appear to move in close correspondence to the number of work units? As described in Section 2, the DFAS price for a specific output equals the expected total cost of producing the expected number of work units, divided by that number of work units. These expected total costs include both fixed costs (those incurred regardless of the number of work units produced) and incremental costs (those that change as output levels change). Assuming DFAS has fixed costs (we believe this is a safe assumption), DWCF-dictated prices will be higher than the incremental production costs.

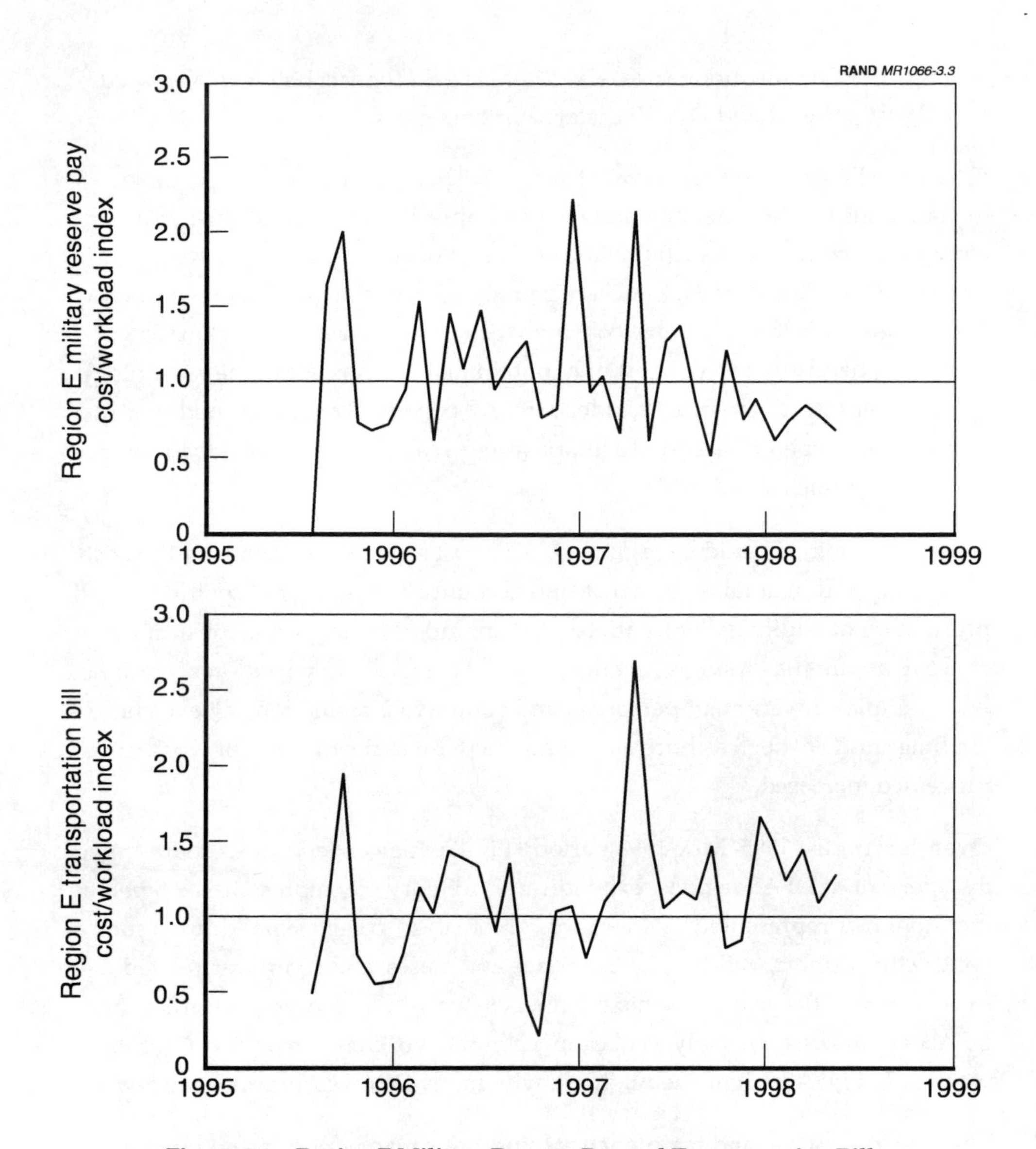

Figure 3.3—Region E Military Reserve Pay and Transportation Bill Cost/Workload Indices

Figures 2.4 and 2.5 suggest that DFAS's incremental costs are considerably less than its average costs. Incremental, or marginal, costs represent how total costs change as output levels change. In the presence of fixed costs, incremental costs can be considerably less than average costs. Reducing workload does not commensurably reduce costs; increasing workload does not commensurably increase costs.[3] Under DWCF rules, however, DFAS revenue rises or falls in

[3]Appendix B provides a detailed set of regression analyses supporting the intuition conveyed in Figures 2.4 and 2.5. In particular, it provides several estimations supportive of the assertion that DFAS's measured efficiency increases when workload increases and conversely, thereby supporting our view that DFAS has considerable fixed costs.

exact proportion to workload levels. There is a disconnect between observed DFAS cost patterns and DWCF-dictated pricing rules.

Why does DFAS have such a cost structure? There are a variety of possible explanations for this phenomenon. For example, it could be that the nature of computer technology is such that a given piece of equipment or software can process 100 or 10,000 travel vouchers equally well, so the incremental cost of a travel voucher is low. Average cost per work unit increases as workload decreases, because the fixed cost of maintaining or depreciating the technology is spread over fewer work units. Indeed, DFAS personnel are concerned that depreciation expenses are to rise in upcoming years, all while workload remains level or continues to fall.

In a similar vein, it could be argued DFAS's cost structure is driven by the costs of creating and maintaining finance and accounting systems, not by the production of additional work units. It is an important and time-consuming chore to assure that systems function without computer bugs and in accordance with complex government personnel and contractual regulations. Nevertheless, the magnitude of such a chore would not increase as the number of work units processed increased.

Over the August 1995–May 1998 period,[4] civilian pay expenditures represented 53.7 percent of DFAS regions' expenditures, military pay represented 2.8 percent, and nonlabor represented 43.4 percent. "Nonlabor" covers expenditures such as contractor support, building maintenance and leases, and computer-related expenditures. Because civilian personnel costs are such a large proportion of DFAS's total costs, an analysis of changes in the workforce structure and total wage bill of DFAS might shed light on why many DFAS costs appear to be fixed.

The system creation and maintenance hypothesis may be supported by Figure 3.4, which shows that DFAS has seen a considerable increase in its relative number of high-grade civilian employees and a diminution in the relative number of low-grade employees. The broken line shows the number of DFAS civilian employees in each civilian General Schedule (GS) grade level in 1993; the solid line is the number of civilian employees in 1996. Between 1993 and 1996, the total number of DFAS civilian employees decreased from 23,561 to 22,001. There was a net reduction in the number of GS-3 through 6 employees, and an increase in the number of higher-cost (and higher-skilled) GS-11 and GS-12 employees. One factor in this grade increase was DFAS's 1994 absorption of

---

[4]We exclude October 1997 from the calculation because we have no data from the Denver region for that month.

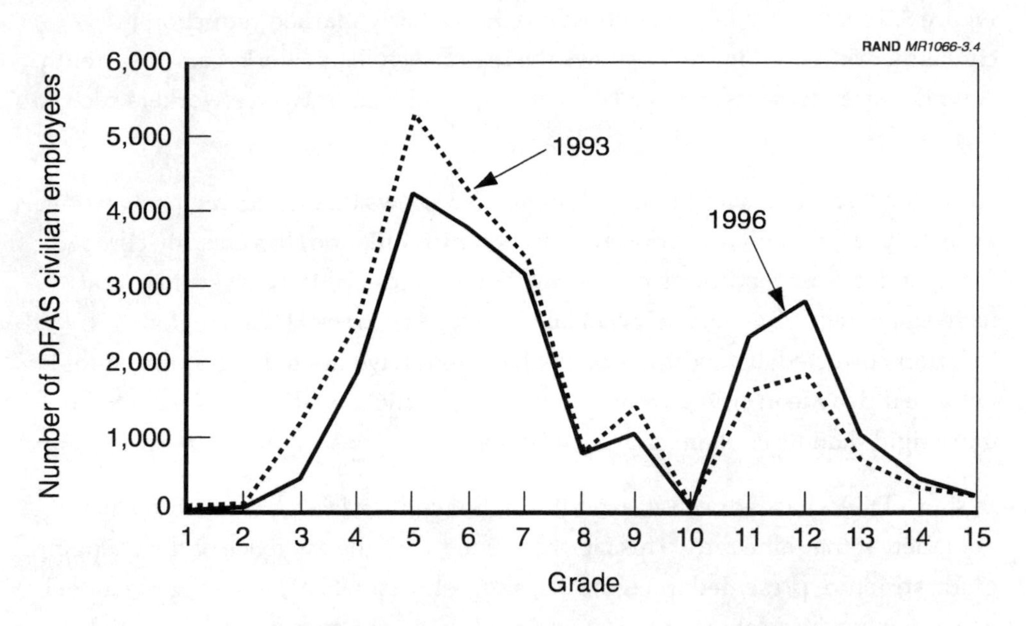

Figure 3.4—DFAS Civilian Grade Structure

high-grade Financial Systems Organization (FSO) personnel who are responsible for systems maintenance functions.[5]

To the extent that systems maintenance activities require more highly skilled employees than work-unit processing activities, this shift might cause DFAS's personnel costs to become proportionally more fixed over time. In other words, more of the personnel are involved in activities that must be done regardless of the number of work units processed, and fewer personnel are involved in the actual processing of work units. For example, DFAS must deal with the Year 2000 (Y2K) problem; the level of effort required to address this problem does not appear to depend on the number of work units processed.

It could also be that DFAS's costs appear to be fixed due to a reluctance or inability to easily reduce civilian employee costs as workload falls. Civil service rules make it difficult to remove a worker without a Reduction in Force (RIF). If a government organization decides to undertake a RIF, there are a variety of complex regulations that must be adhered to. See Robbert, Gates, and Elliott (1997). RIFs draw considerable attention. DFAS managers are not able to add or subtract workers as they see fit.

---

[5]These data come from civilian personnel data files maintained by the Defense Manpower Data Center (DMDC).

However, DFAS has been able to reduce its workforce, mostly by attrition. Figure 3.5 shows that DFAS regions have had a fairly marked reduction in civilian workforce. Figure 3.5 shows the regions' civilian "workyears" by month. A workyear represents one worker working a full year or twelve workers each working one month.

However, in accord with Figure 3.4, Figure 3.6 shows that DFAS regions' average expenditure per civilian workyear (correcting for inflation) has steadily climbed. Rising average expenditures per workyear have reduced the cost savings that have emanated from DFAS civilian labor force reductions. DFAS regions' inflation-corrected expenditures on civilian labor have not fallen commensurably with their decline in civilian workyears. DFAS's civilian labor costs seem to be more rigid than its civilian labor force, per se.

In short, DFAS has reduced its labor force, but civilian labor expenditures have not fallen at the same rate. This fact, combined with the evidence of the changing grade structure presented in Figure 3.4, suggests that DFAS is making personnel adjustments where labor input is workload dependent, but that a substantial portion of the labor effort is not workload dependent.[6]

---

[6]Robbert, Gates, and Elliott (1997) note that civil service layoff procedures tend to result in mostly junior, low-grade personnel leaving federal employment. Even if high-grade positions are eliminated, civil service employees have bumping and retreating rights that can result in the release of employees at a lower grade than those holding the eliminated positions. Also, tactics like hiring freezes tend to result in rising average grade levels as remaining personnel progress through the grade structure. However, such tactics, while increasing the average grade (as shown by the increasing average labor costs in Figure 3.6), should not increase the absolute number of higher-graded personnel (Figure 3.4).

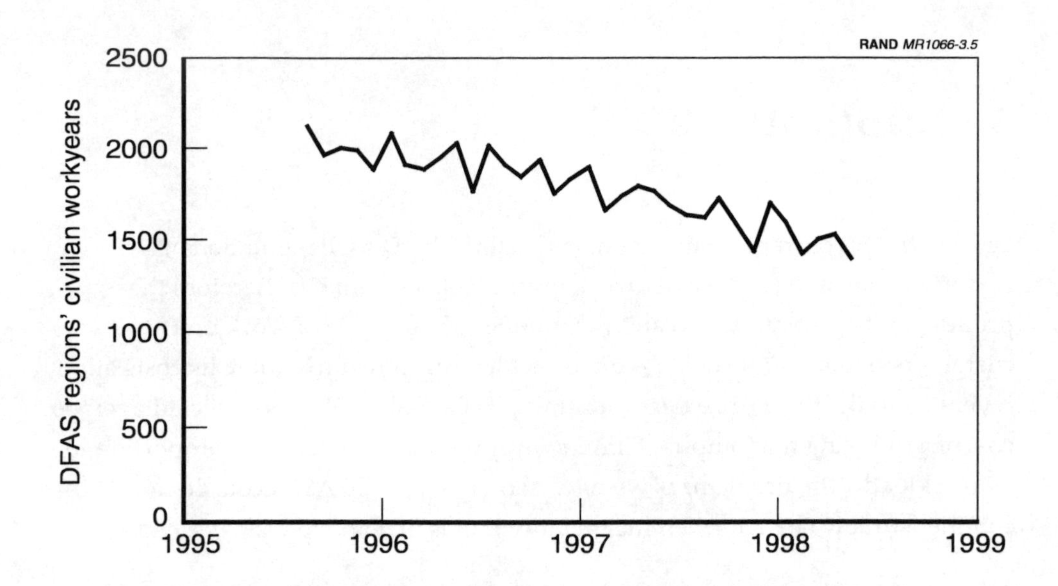

Figure 3.5—DFAS Regions' Civilian Workyears

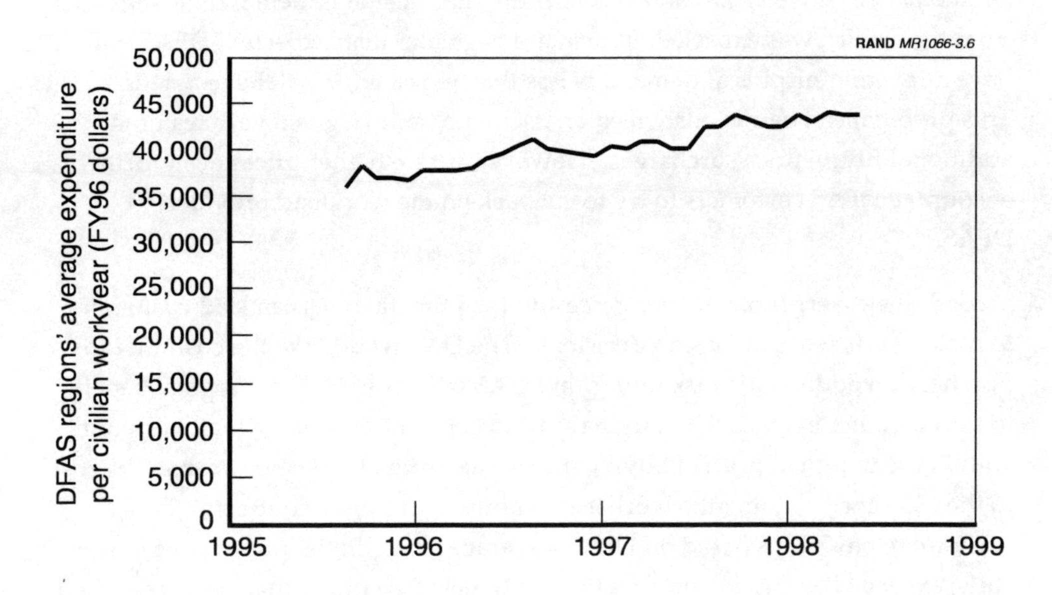

Figure 3.6—DFAS Regions' Civilian Labor Expenditures per Workyear

4. Implications

Under DFAS's current arrangement (as dictated by DWCF regulations), customers pay a fixed price per work unit; DFAS's revenue is therefore the product of its price vector (a) and the number of each type of work unit customers demand (Q). DFAS's price vector is supposed to equate its costs and revenues so the fixed price per work unit is set equal to DFAS's expected average cost of producing that output. DFAS's revenue changes exactly in proportion to its workload. The problem, as we have shown, is that DFAS's costs do not commensurably change when the quantity that is demanded changes.

This divergence between price and incremental costs creates two important problems within the existing DWCF system. First, as mentioned earlier, there is an imbalance between cost and revenue any time customer demand deviates from that which was expected. If demand is greater than expected, DFAS will have a revenue surplus; if demand is less than expected, it will have a deficit. This problem can perniciously feed on itself. Losses in a given year result in additional future price surcharges. However, future higher prices even further encourage DFAS customers to try to cut back on the workload provided to DFAS.

Second, the divergence between price and incremental costs can lead customers to make inefficient purchasing decisions. The DoD would like the customer to purchase an additional work unit from DFAS whenever (1) the marginal benefit to the customer exceeds the marginal cost to DFAS of producing that work unit and (2) the marginal price of buying the service from DFAS is lower than the cost to the customer of doing the work itself or buying it from a contractor. Customers pay prices based on DFAS's average costs. In the presence of fixed costs, average cost pricing implies that customers face prices that are higher than the incremental cost of producing the good. As a result, customers may decide to consume less than the optimal amount, or they may try to produce it themselves or hire a contractor. Such substitution is often grossly inefficient from an organizational perspective. Average cost pricing provides customers with the wrong incentives.[1]

---

[1]It is an open question to what extent DFAS customers are responsive to DFAS prices. For example, we have been told the Army has different people requesting work from DFAS (e.g., corps comptrollers) from those who pay for such work (Army headquarters). Such a consumer-payer disconnect would tend to lessen customer price responsiveness. Of course, DFAS cannot control how

We believe DFAS's current pricing structure is unsustainable. In particular, as long as demand for DFAS's services continues to fall faster than planned, the current DWCF pricing rules imply DFAS's revenues will chronically fall short of DFAS's costs. Some marked change in DFAS's pricing structure is needed.

How might DFAS move toward a more appropriate pricing strategy? A private-sector firm faced with such a cost structure could try a variety of tactics. A common approach is to establish a nonlinear pricing schedule. A nonlinear pricing schedule typically involves some sort of up-front fee, followed by lower incremental charges for additional units. Often, a customer facing a nonlinear price schedule gets a considerable discount for larger quantities.

A classic example of nonlinear pricing schemes is found in amusement parks, which often charge customers an entry fee and then an additional cost per ride (Oi, 1971). The telecommunications industry, which faces very high fixed costs and very low incremental costs, also has many examples of nonlinear pricing schemes. A digital phone service package might charge a flat fee of $100 per month for the basic service that includes 1000 "free" minutes of calling anywhere in the United States, and then charge $.10 per minute for minutes over 1000. People who actually use 1000 minutes or more per month pay $.10 per minute, whereas people who use less pay more per minute. Other plans allow users "unlimited" calling on nights and weekends, but charge a high per-minute rate during business hours.

These are just a few examples of a wide array of pricing options employed by profit-maximizing firms that face large fixed costs. Related approaches include bundling products with different cost structures and nonuniform price schedules, where different groups of customers are charged different prices and/or fees. An example of a nonuniform price schedule is Ramsey pricing, where customers who are less sensitive to price changes are charged more. Appendix C discusses the academic literature on nonlinear pricing in more depth.

One possible approach for DFAS would be to establish prices equal to its estimate of long-run incremental costs of each output. It is unlikely that long-run incremental costs would be equal to average costs. If DoD were interested in balancing the DFAS budget, it would also need to establish lump-sum transfer payments from DFAS customers to DFAS. This could be accomplished through a simple nonlinear pricing structure.

---

customers set up their internal arrangements and must be prepared to possibly face more price responsive customers in the future.

Under a simple nonlinear pricing structure, DFAS customers would make fixed annual payments (*b*) to DFAS for system creation and maintenance, plus smaller-than-current incremental payments that vary based on the quantity of various outputs demanded (cQ). The incremental payment rate (c) would be set equal to DFAS's incremental cost of producing such work units.

Such a pricing structure would better align incremental customer prices with incremental DFAS costs, thereby allowing customers to make better decisions about what level of DFAS services to consume. This proposal is similar to the Air Force pricing proposal made by Baldwin and Gotz (1998). Their view is that Air Force depot-level reparable prices should be the depots' expected marginal costs of such repairs, and that overhead should be recovered through separate charges to customer commands. The Defense Science Board 1996 Summer Study also urged that fixed costs not be included in revolving fund charges.

Of course, a $b + cQ$ pricing schedule is just one possible alternative fee structure. As shown in Appendix B, DFAS's actual cost structure appears to be more nonlinear than that. However, a simple nonlinear pricing schedule may initially be preferable to a more complex approach, both administratively and politically. DFAS may have to experiment to determine the most appropriate nonlinear pricing schedule.

The DFAS price structure is based on average costs across all DFAS customers, with one exception. The RADSS data suggest that different DFAS customers impose different costs on DFAS for the same output. For example, different regions' average costs per travel voucher work unit vary by more than a factor of 4. The same sort of pattern holds for other outputs. As noted above, regions tend to work for specific service customers. Figure 4.1 shows that different regions have very different average costs per travel voucher work unit.

These differences in average cost across regions for a given output may stem from differences in incremental and/or fixed costs, different customers impose different processing burdens on DFAS. Figure 4.1 suggests it might be most appropriate to have a nonlinear price schedule such as $b_i + c_iQ_i$, where the fixed and variable fees vary across customers.

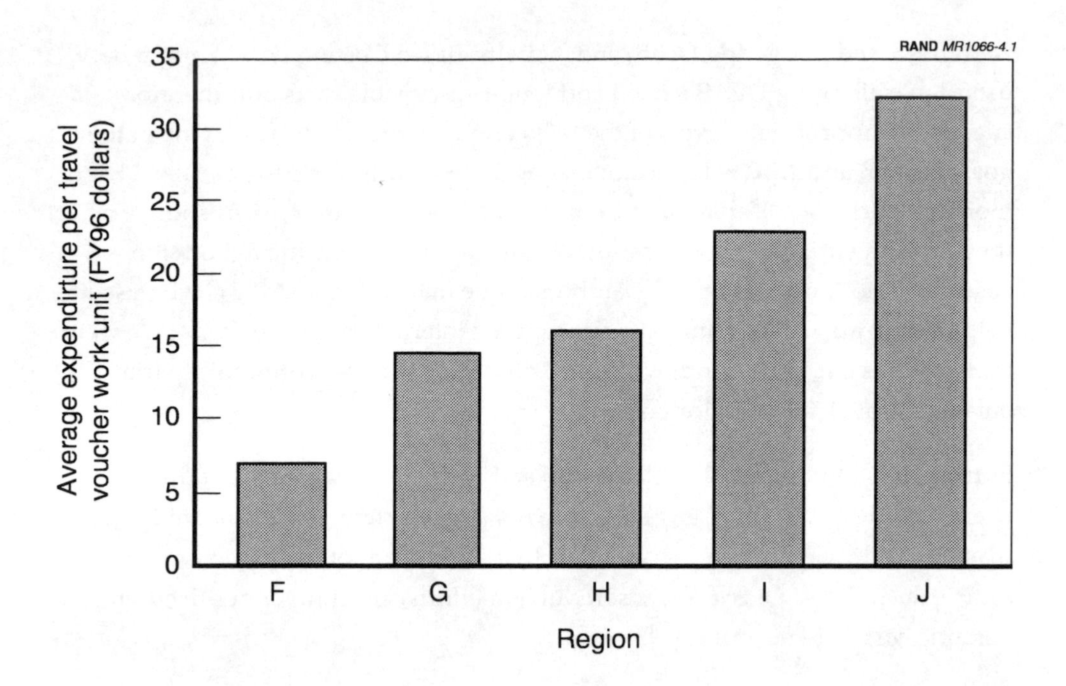

Figure 4.1—Regions' Average Costs per Travel Voucher Work Unit

DFAS already charges different customers different fees for monthly trial balance processing.[2] That fee arrangement, however, is linear (a_iQ_i). All other outputs have the same price per work unit for every customer (aQ_i).

We do not wish to understate the challenges associated with changing DFAS's pricing algorithm. For example, careful analysis would need to be undertaken to determine how best to divide DFAS's fixed costs among its customers. Low-volume customers, such as the Marines, may pose a particular challenge; the current linear pricing system may be disproportionately favorable toward them. Cost analysis would also be needed to determine appropriate incremental fees.

Although economic and accounting literature suggests an organization set transfer prices equal to long run incremental costs when competitive market prices are unavailable, it provides little guidance as to how to determine long run incremental costs. The quality of such a revised pricing system will be directly related to the quality of the cost information gathered. It is worth stressing, however, that even less-than-perfect information about incremental costs will likely generate more effective prices than those based on average costs.

---

[2]We were informed that DFAS plans to go to per-labor-hour charges for monthly trial balance work, starting in October 1999.

Activity-Based Budgeting (ABB) and Activity-Based Costing (ABC) might be useful in calibrating DFAS's fixed and long-run variable costs and therefore suggesting appropriate levels of the fixed and incremental prices.[3] One of the goals of ABB and ABC is to induce managers to view fewer costs as "fixed" and more as "variable." Kaplan and Cooper (1998b) suggest that prices should reflect the long-run variable cost of production and should be specific to both the customer and the product. The authors argue that ABB and ABC can be used to help organizations determine the appropriate internal transfer prices. (We were told DFAS is currently conducting ABC studies in the areas of monthly trial balances, travel, and vendor pay.)

Further, it is important to recall that "fixed" costs are only those that do not change as output levels change. We are not suggesting that the current level of fixed costs should not be questioned or that efforts cannot be undertaken to reduce such costs. Of course, cost reduction efforts, affecting either fixed or variable costs, are never easy.[4]

---

[3]Cooper and Kaplan (1988, 1991, 1998a) have a set of articles on this topic. See Appendix D for a discussion of ABC and ABB.

[4]Appendix E details DFAS's limited experience with attempting to reduce costs through the A-76 cost comparison process.

5. Conclusions and Recommendations

Conclusions

This report has presented an analysis of the costs and performance of the DFAS. The data show that while DFAS performance has improved, it is limited by DWCF-dictated pricing policies. Under current DWCF rules, the price per work unit is based on expected average costs and expected workload. With this linear pricing structure, DFAS revenue increases and decreases in exact proportion to actual DFAS workload. When the actual workload is less than the expected workload, DWCF policies may lock DFAS into losses.

A major and recurrent finding of our analysis is that many DFAS costs do not vary in direct proportion to DFAS workload levels. These costs may be relatively fixed for observable, technology-related reasons, or because of an inability to pursue policies that would generate sufficient cost reductions. This nonlinear nature of DFAS costs is in direct contrast with DWCF pricing policies. The linear DWCF-dictated prices may not provide adequate incentive for customers to use DFAS services at the expected workload level, thus leading to inadequate revenue. Prices must then be raised to make up for revenue losses, and/or costs must be reduced. Higher prices can further impact workload levels. Cost reductions have not kept pace with revenue decreases, and it may be difficult for them ever to do so.

We believe DFAS's current pricing structure is unsustainable. The pricing policies that DFAS uses need to be changed to account for the reality of cost rigidities. We strongly suspect that this holds true for other DoD and governmental entities.

Recommendations

Our recommendation is that DFAS experiment with a simple nonlinear pricing structure. Under such an arrangement, DFAS would receive "open the door" transfer payments from its customers that reflect DFAS's fixed costs. DFAS would also assess incremental fees per unit of workload, which would be lower than current DFAS prices. Such a structure would give DFAS customers more-appropriate incentives with respect to how much workload to give DFAS. Both transfer and incremental fees could vary across customers as well as outputs.

It would be more straightforward to get approval for DFAS to deviate from DWCF pricing policies for this experimentation, rather than to attempt to change DWCF pricing policies themselves. Successful experimentation may then lead to implementation of nonlinear pricing at other DWCF entities with similar cost structures.

We do not wish to understate the challenges associated with changing DFAS's pricing algorithm. Careful analysis would need to be undertaken to determine how best to divide DFAS's fixed costs among its customers. (While acknowledging some DFAS costs to be invariant to output, or fixed, efforts should continue to reduce these costs.) Similarly, cost analysis would be needed to determine appropriate incremental fees.

Appendix

A. DFAS Cost/Workload Performance Time Trends

This appendix presents regression results supporting the findings in Section 3 of this report. Specifically, in Section 3, we make a number of assertions about performance trends. In this appendix, we undertake a series of simple linear regressions to show that the trends we assert in the data do appear to exist, if one accepts the assumptions necessary to validly utilize linear regression. The purpose of this estimation is description of historical performance, not an attempt to extrapolate to estimates of future performance.

We analyzed the statistical significance of Figure 3.1's downward trend by regressing each month's cost/workload index value on the month's numerical value (e.g., July 1997 is 1997+6.5/12; the midpoint of the month of July is 6.5/12 through the calendar year). Table A.1 gives the regression results.

Though the R-squared statistic of this regression is poor, we see there is a negative (favorable), statistically significant time trend in Figure 3.1's data. To put these coefficient estimates in perspective, the regression estimates a fitted January 1996 (1996+0.5/12) cost/workload index of about 1.04. The month coefficient estimate of –0.06 suggests the January 1997 fitted cost/workload index

Table A.1

Aggregate DFAS Regions' Cost/Workload Index Time Trend Analysis

Observations	33
R-squared	0.14

	DF	SS
Regression	1	0.07
Residual	31	0.44
Total	32	0.51

	Coefficient	SE	T statistic	P-Value
Intercept	114.60	50.72	2.26	0.03
Month	–0.06	0.03	–2.24	0.03

is about 0.98. Of course, we are not suggesting such improvement will continue indefinitely. We are merely showing that a pattern of improvement is observable in these historical data.

Figure A.1 graphically presents Table A.1's regression results. We depict the apparent trend line along with the actual data.

We are concerned, however, that a few data (e.g., August 1995's poor performance and May 1998's good performance) are causing the improved performance finding. Through most of the data, performance has been largely static. As shown in Table A.2, there is no significant time trend if the August 1995 and May 1998 data are excluded.

Table A.3 shows a regression analysis of Region C's cost/workload trend, shown in Figure 3.2. There appears to have been a trend of improvement, as shown by the "Month" coefficient estimate of –0.17.

Table A.4, meanwhile, shows no time trend in Region D over the same period. The "Month" coefficient is insignificantly different from 0. Tables A.3 and A.4 corroborate our observations about Figure 3.2.

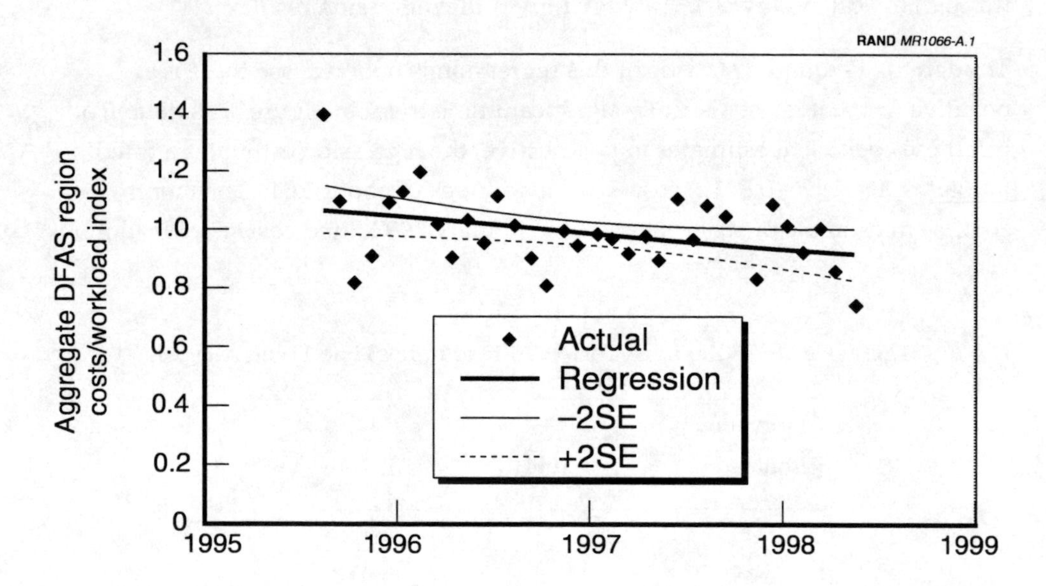

Figure A.1—Table A.1 Regression Results

Table A.2

Aggregate DFAS Regions' Cost/Workload Index Time Trend Analysis Excluding August 1995 and May 1998

Observations	31	
R-squared	0.02	

	DF	SS
Regression	1	0.01
Residual	29	0.28
Total	30	0.28

	Coefficient	SE	T statistic	P-Value
Intercept	39.92	45.66	0.87	0.39
Month	–0.02	0.02	–0.85	0.40

Table A.3

Region C Cost/Workload Index Time Trend Analysis

Observations	43	
R-squared	0.27	

	DF	SS
Regression	1	1.36
Residual	41	3.68
Total	42	5.05

	Coefficient	SE	T statistic	P-Value
Intercept	342.64	87.65	3.91	0.0003
Month	–0.17	0.04	–3.90	0.0004

For Figure 3.3, excluding the July 1995 negative expenditure outlier, Table A.5 shows that Region E's military reserve pay account performance appears to improve (though this trend only exists at the 90 percent confidence level—the P-value is less than 0.1, but greater than 0.05).

Table A.6 shows that Region E's transportation bill performance appears to have worsened (at the 90 percent confidence level). Note the "Month" coefficient estimate is positive/unfavorable in this regression.

Table A.4

Region D Cost/Workload Index Time Trend Analysis

Observations	44	
R-squared	0.002	

	DF	SS
Regression	1	0.003
Residual	42	1.516
Total	43	1.518

	Coefficient	SE	T statistic	P-Value
Intercept	15.60	53.05	0.29	0.77
Month	–0.01	0.03	–0.27	0.78

Table A.5

Region E Military Reserve Pay Account Cost/Workload Index Time Trend Analysis

Observations	34	
R-squared	0.10	

	DF	SS
Regression	1	0.64
Residual	32	5.82
Total	33	6.46

	Coefficient	SE	T statistic	P-Value
Intercept	337.00	178.59	1.89	0.07
Month	–0.17	0.09	–1.88	0.07

Table A.6

Region E Transportation Bill Cost/Workload Index Time Trend Analysis

Observations	35	
R-squared	0.09	

	DF	SS
Regression	1	0.65
Residual	33	6.61
Total	34	7.26

	Coefficient	SE	T statistic	P-Value
Intercept	–321.74	179.54	–1.79	0.08
Month	0.16	0.09	1.80	0.08

B. Regression Evidence of DFAS's Cost Rigidity

This appendix presents regression analysis underlying the fixed-cost findings in Section 3. In particular, the focus of this appendix is to lend support to the assertion in Section 3 that DFAS costs do not change in proportion to DFAS workload. Put differently, Section 3 asserts that DFAS appears to have fixed costs and that increasing DFAS workload improves DFAS efficiency as we measure it.

Pooling together all our regions' and outputs' data, we regressed each region's and output's cost/workload index on the month value (to assess a time trend) and on what we term the month's "work ratio." A month had a work ratio of 1.0 if the output's work unit level that month equaled the average monthly work unit level for that output in that region. By contrast, a work ratio of 0.5 indicates the region had only half its typical work unit level for that output in that month.

As shown in Table B.1, the coefficient estimates on both the month value (–0.06) and the work ratio value (–0.86) are negative and significant (P-values substantively below 0.01). The month estimate suggests DFAS regional performance typically improved through the data, controlling for generally falling workload. The work ratio estimate corroborates Figures 2.4 and 2.5 in suggesting that cost/workload index–measured performance improved, i.e., the index fell, when workload rose. Put differently, efficiency tended to worsen when workload fell. This result is consistent with DFAS regions having considerable fixed costs.

We extended Table B.1's results by computing separate "work ratio" estimates for each separate output. We are testing the hypothesis that different outputs might have different fixed-cost characteristics. As shown in Table B.2, all DFAS outputs show a strong tendency for costs to remain more stable than workload. All the outputs have significantly negative ratio coefficients. The phenomenon asserted in Figures 2.4 and 2.5 appears to hold for all DFAS outputs.

Table B.3 extends Table B.1's results further by including the square of the workload ratio as an independent variable. This structure allows for nonlinear effects of workload on the cost/workload index. Under this parameterization,

Table B.1

DFAS Region Regression of Monthly Cost/Workload Indices on Time Trend, Workload Level

Observations	1515
R-squared	0.21

	DF	SS
Regression	2	156.60
Residual	1512	586.19
Total	1514	742.79

	Coefficient	SE	T statistic	P-Value
Intercept	116.85	33.74	3.46	0.0005
Month	–0.06	0.02	–3.40	0.0007
Work Ratio	–0.86	0.04	–19.30	0.0000

we find the work ratio coefficient is more negative than in Table B.1, but the square of the work ratio has a positive coefficient estimate. These parameter estimates are consistent with the argument that increasing workload increases efficiency for a while, but eventually too much workload would overwhelm the system.

Figure B.1 plots Table B.1's and Table B.3's fitted values as a function of the work ratio. Figure B.1 shows that Tables B.1 and B.3 provide only subtly different conclusions within the typical region of workload values (halving to doubling of output levels). Both parameterizations strongly suggest that measured performance improves considerably (the estimated cost/workload index falls) as workload levels increase. Our DFAS fixed-cost finding appears to be robust to different estimation approaches.

Table B.2

DFAS Region Regression of Monthly Cost/Workload Indices on Time Trend, Output-Specific Workload Level

Observations	1515
R-squared	0.24

	DF	SS
Regression	15	175.91
Residual	1499	566.88
Total	1514	742.79

	Coefficient	SE	T statistic	P-Value
Intercept	116.93	33.45	3.50	0.0005
Month	–0.06	0.02	–3.43	0.0006
Civilian Pay Work Ratio	–0.85	0.06	–14.16	0.0000
Civilian Pay Part Work Ratio	–1.14	0.14	–8.27	0.0000
Commercial Invoice Work Ratio	–1.01	0.06	–16.03	0.0000
Contract Invoices (MOCAS) Work Ratio	–1.06	0.12	–9.11	0.0000
Contract Invoices (SAMMS) Work Ratio	–0.90	0.11	–8.49	0.0000
Finance & Accounting Commissary Work Ratio	–1.04	0.11	–9.07	0.0000
Military Active Pay Work Ratio	–1.08	0.07	–15.45	0.0000
Military Reserve Pay Work Ratio	–1.08	0.07	–15.41	0.0000
Military Retired Pay Work Ratio	–1.06	0.08	–12.94	0.0000
Military Pay Incremental Work Ratio	–1.07	0.12	–9.28	0.0000
Monthly Trial Balance Work Ratio	–1.04	0.06	–16.05	0.0000
Out of Service Debt Work Ratio	–0.75	0.05	–13.98	0.0000
Transportation Bill Work Ratio	–0.89	0.07	–13.73	0.0000
Travel Voucher Work Ratio	–0.96	0.06	–15.39	0.0000

Table B.3

DFAS Region Regression of Monthly Cost/Workload Indices on Time Trend, Workload Level, Square of Workload Level

Observations	1515
R-squared	0.31

	DF	SS
Regression	3	230.69
Residual	1511	512.10
Total	1514	742.79

	Coefficient	SE	T statistic	P-Value
Intercept	109.90	31.55	3.48	0.0005
Month	–0.05	0.02	–3.39	0.0007
Work Ratio	–2.13	0.10	–22.31	0.0000
Work Ratio Squared	0.40	0.03	4.79	0.0000

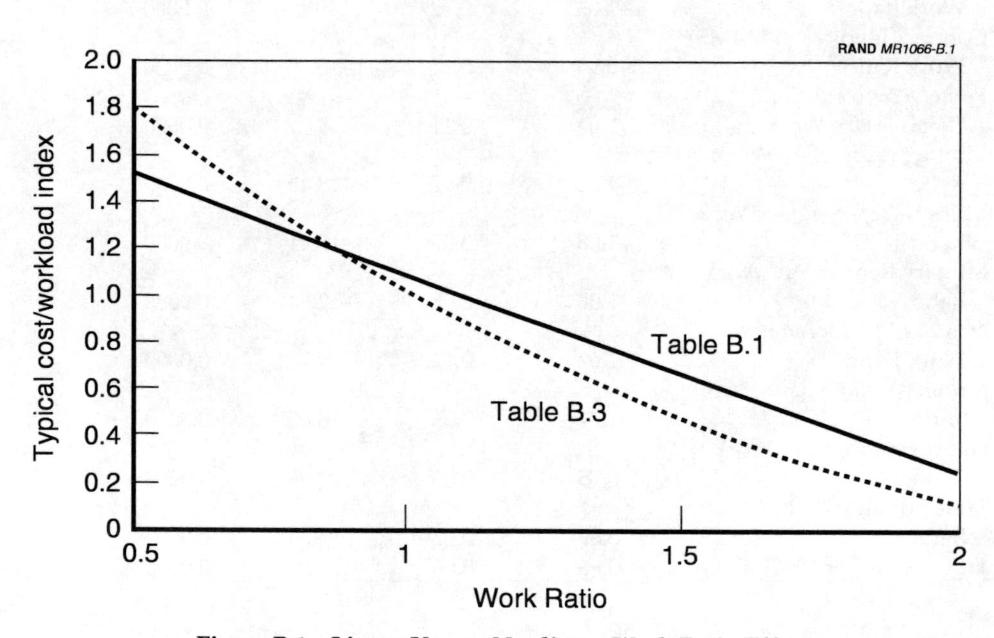

Figure B.1—Linear Versus Nonlinear Work Ratio Effects

C. The Academic Literature on Nonlinear Pricing

Nonlinear pricing arrangements can have positive or negative implications for social welfare,[1] depending on the competitive environment, the objective of the firm, the specific pricing options employed, and the demand characteristics of the customers. For example, a nonlinear pricing structure involving a high up-front fee and a low usage cost might reduce social welfare if the fee is too high for a lot of potential customers. However, such a pricing structure might actually increase social welfare relative to a flat per-unit fee if it allows the producer to sell to more customers or serve additional markets. Economists have shown that, in general, a profit-maximizing nonuniform price schedule does not maximize social welfare but that nonuniform and/or nonlinear price structures can improve social welfare relative to a single fixed price under certain conditions (Spence, 1977; Katz, 1983; Varian, 1985). In order for such pricing structures to improve social welfare, it must be the case that the firm reaches a new market or new customers through the alternative pricing structure. Hausman and Mackie-Mason (1988) show that when a firm faces decreasing marginal costs, nonuniform price structures can improve social welfare even when the firm is not reaching a new market.

Many of the potentially negative implications of nonlinear pricing stem from the "greedy" behavior of profit-maximizing firms. Firms interested in maximizing profits do not care if they cut certain consumers out of the market and harm social welfare as long as they can increase their own profits. Even in the for-profit world, such a scenario is rare, applying mainly to monopolies. For DFAS, these problems are even less likely because the problem faced by the Department of Defense in determining an effective pricing arrangement for DFAS is fundamentally different from the problem faced by a profit-maximizing firm. In fact, it is more akin to what economists call a "social planning problem" or an

[1]"Social welfare" is a general term that refers to the total net benefit that all relevant parties (e.g. customers, producers, workers, innocent bystanders who might by impacted) derive from the activity in question. In the case of DFAS, social welfare refers to the total value that the customer organizations (e.g., the installations or DoD) derive from DFAS services and the total benefit that DFAS and its workers derive from producing the output. Because social welfare considers the total benefit derived by both the producer and consumer, it is not influenced directly by payments from the customer to the producer. However, payment arrangements can influence social welfare indirectly through the amount of output the customer demands.

"internal transfer pricing problem."[2] The distinction is that the DoD is interested not in maximizing the profitability of DFAS but in achieving the most efficient use of resources.

In the context of the social planning problem, a planner who has perfect information on the benefit function of the consumer and the cost function of the producer can direct an efficient level of production.

The problem is that the planner normally does not have perfect information about cost and benefit functions. One reason for this informational deficiency is that economic conditions change between the time the plan is set and the time it is implemented, altering cost and benefit functions. Another problem is that production managers have more information about the production process than the government or owner of the company. However, they also have different objectives, and as a result, they cannot be relied upon to use that information in the same way that the government or owner would. As a result, the social planning or transfer pricing[3] problem is one of asymmetric information and divergence of preferences.[4]

Much has been written about how transfer prices should be determined in order to promote the optimal transfer of resources within an organization under these conditions.[5] When the good or service is offered in perfectly competitive markets, economists advise that an organization use that market price. When the good or service is not exchanged in perfectly competitive markets, then the transfer price should be set equal to the long-run marginal cost of production.[6] Then, a planner interested in balancing the budgets of the units involved in the transfer can specify a lump-sum transfer in addition to the incremental prices. It is important to note, however, that the incremental prices are set to promote the

---

[2]See, for example, Malinvaud (1967) and Williamson (1967) on the social planning problem and Harris, Kriebel, and Raviv (1982) on the transfer pricing problem. The social planning problem is slightly more general in that it allows the planner to set production quotas rather than prices if that is optimal. The transfer pricing problem restricts the planner (normally a firm owner) to set a price and then allow managers to determine the quantity of goods exchanged at that price. Weitzman (1974) points out that under perfect information, the planner can achieve the same results by setting a per-unit price equal to the marginal cost of producing the output and letting the producer and consumer determine the appropriate number of units exchanged, or by setting the optimal level of production and letting the producer establish the price. Under perfect information, the two approaches are equivalent.

[3]In the economic literature, this problem is often described as the transfer pricing problem because the focus is on private-sector companies in which the notion of setting transfer prices is viewed as superior to setting quantities.

[4]See, for example, Holmstrom and Tirole (1991) and Harris, Kriebel, and Raviv (1982).

[5]See, for example, Eccles (1985), Eccles and White (1988), and Baldwin and Gotz (1998) for reviews of transfer pricing.

[6]See Hirschleifer (1956) and Solomons (1965).

efficient use of resources, whereas any lump-sum transfer serves purely an accounting function.

D. Activity Based Costing (ABC), Activity Based Budgeting (ABB), and Implications for Public Sector Organizations

In a series of articles and books, Cooper and Kaplan (1988, 1991, 1998a, 1998b) have developed a new perspective on accounting and budgeting, referred to as Activity Based Costing (ABC) and Activity Based Budgeting (ABB). This work has generated a revolution in the business world in term of the way cost information is collected and used.

Activity Based Costing

Activity Based Costing is motivated by a belief that traditional (general ledger) accounting information is all but useless to managers who are interested in evaluating the effectiveness of resource allocation decisions in their companies. This traditional information is geared instead toward satisfying auditors or other outsiders who are interested in some evidence of financial accountability. According to Cooper and Kaplan (1988), one of the most serious problems lies in the traditional overhead cost-allocation process. Over time, as production processes have become more and more complex, a greater proportion of total production costs are described as "overhead" and are arbitrarily allocated to output. The authors suggest that many of these "overhead costs" (e.g., costs of logistics, production, marketing, sales, distribution, service, technology, financial administration, information resources, and general administration) can, in fact, be traced to individual products or product groups. Certain activities and processes consume a disproportionate amount of these activities. Cooper and Kaplan argue that the misallocation of overhead costs can generate tremendous distortions in production cost estimates. Specifically, traditional costing strategies tend to attribute too much overhead to less-complex products and products produced in high volume. Conversely, they seriously underestimate low-volume, complex products and services. Because this cost information is often used to evaluate the profitability of different production strategies, the misallocation of costs can lead managers to make poor decisions.

Cooper and Kaplan propose ABC as an alternative to these traditional accounting systems. Under ABC, the production process is viewed as a set of activities. Managers are asked to consider the resources consumed by these different activities, and only then to assign activities to products and/or customers. Having considered the relationship between an activity and a product or customer, indirect costs can be more appropriately assigned to those products or customers. To put it simply, you have to know what needs to be done to produce a product before you can figure out how much that product costs. Cooper and Kaplan (1991) distinguish between four basic types of activities:

1. Facility-sustaining activities (utilities, building and grounds, plant management)
2. Product-sustaining activities (process engineering, product specs, product enhancement)
3. Batch-level activities (setups, material movements, purchase orders, inspections).
4. Unit-level activities (direct-labor, materials, machine costs, energy).

Cooper and Kaplan argue that breaking down costs in this way can induce managers to consider a wider array of cost-saving strategies. Heretofore, most cost-cutting efforts have focused on the unit-level activities because those costs were most visible. They argue that there are significant opportunities for cost savings in batch-level and product-sustaining activities.

In undertaking an Activity Based Costing analysis, Cooper and Kaplan caution that managers must carefully distinguish the costs that fall in each category and refrain from allocating facility, product-sustaining and batch-level costs to individual units. In particular, facility-sustaining costs should *not* be assigned to individual products.

The authors also stress that managers need to consider the cost of excess capacity as a separate line item, rather than wrapping it up into an estimate of incremental costs. This is because the cost of excess capacity does not reflect anything about the productivity of the capital or labor. Calculating per-unit costs on the basis of product volume can lead to a "death spiral": Because it looks like per-unit costs are rising dramatically when volume declines, management raises prices, higher prices lead to further volume declines, and these volume declines lead to further price increases.

Activity Based Costing is a not a trivial endeavor. It is an entirely new cost system requiring a new way of thinking about costs. In most cases, ABC must be implemented in addition to the traditional cost systems, which are required for accountability purposes. Once an organization decides to implement ABC, it must determine the level of detail it wishes to collect, recognizing that more precise information is much more costly to collect. At some point the additional detail is not worthwhile. Cokins, Stratton, and Helbling (1992) provide a useful implementation-focused overview of ABC and discuss these trade-offs. For example, organizations might want to focus on particularly expensive resources, on resources whose consumption varies by product, or on resources whose demand patterns are not correlated with the traditional allocation measures. DFAS might wish to first focus on its highest dollar outputs, e.g., monthly trial balances.

Activity Based Budgeting

Once an organization has an ABC system in place, it can use ABC information in its budgeting process. Activity Based Budgeting is described by Cooper and Kaplan as "ABC in reverse." The need for ABB is motivated by the observation that the traditional budgeting process in organizations is a negotiation between managers and senior executives over some small percentage change relative to last year's budget and that it rarely revisits issues such as productivity and the effective use of resources. With ABB, managers are induced to consider what resources are actually needed. First, managers develop an estimate of the production and sales volume for the next period. Then, they forecast demand for activities within the organization. They then calculate the demand for resources stemming from those required activities. The next step is to determine the actual resource supply based on spending patterns and the activity capacity. The activity capacity may differ from estimated production volume because some resources are lumpy (i.e., you might only need 1.2 trucks but you have to purchase two because you can't buy a fraction of a truck).

ABB forces managers to think of more fixed costs as variable in the medium to long run. Cooper and Kaplan (1998a, p. 13) "prefer to use the term *committed cost*, since managers have committed the supply of resources in advance and will not alter their supply, in the short run, because of short term demand fluctuations." In other words, fixed costs are fixed because of management decisions, and managers have the flexibility to redeploy those resources as conditions change.[1] Kaplan and Cooper (1998b, p. 302) note

[1]As noted above, DFAS behaves as if many of its costs are fixed. It is unclear whether this apparent rigidity emanates from management decisions, political inflexibility, technological factors, or all of these reasons.

> Left unaddressed by conventional variable or marginal cost thinking is the entire organizational infrastructure of 1) personnel—front-line employees, engineers, salespersons, managers—with whom the organization has a long-term contractual commitment, 2) equipment and facilities, and 3) information systems supplying computing and telecommunications. Decisions to acquire new resources or to continue to maintain the current level of these committed resources are most likely made during the annual budgeting process. Once the authorization to acquire and maintain organizational resources has been made, the expenses of these resources appears fixed and unrelated to local, short term decisions about product mix pricing, and customer relationships.

Kaplan and Cooper (1998b) also advocate the use of ABC for the determination of transfer prices in an organization. They describe the problem of a pharmaceutical company that was using unit-level marginal costs as the transfer price for drugs between the marketing and production units. The problem was that costs were highly sensitive to batch sizes, but the incremental pricing strategy did not reflect this batch size sensitivity. As a result, marketers were selling drugs in inefficiently small batches because they were not receiving the proper signals about the impact that their sales practices were having on cost. Kaplan and Cooper argue that transfer pricing policies should acknowledge the difference between different types of costs. Unit costs should be charged on the basis of quantity consumed. Customers should be charged for batch-level costs on the basis of how many batches are involved in producing the output for them. Finally, product-sustaining and facility-sustaining costs should be charged on the basis of budgeted information, possibly through a fixed payment rather than a per-unit charge.

Many of these cost-allocation and budgeting issues are particularly salient for public sector organizations. Improved cost information is particularly important for organizations interested in outsourcing and privatization (GAO, March 1997). In the federal government, outsourcing is governed by the Office of Management and Budget's Circular A-76. The A-76 circular specifies the terms and conditions for private-public cost comparisons. In 1996, OMB revised the guidelines so that the public sector bid would be required to include a 12 percent overhead charge. The GAO (February 1998, p. 9) has expressed concern that there is no empirical basis for the 12 percent rate and suggested that agencies begin to collect information that could help them identify these costs. "Activity Based Costing is an analytical tool that can be used, generally in conjunction with existing accounting systems, to identify all costs—both direct and indirect—of providing a service or performing a function."

The city of Indianapolis has been on the forefront of implementing ABC. Officials there reported that they were able to develop ABC information by massaging their existing cost data, with the help of a private sector consulting firm. Goldsmith (1997, p. 112–113) reports that in Indianapolis, cost information can have an important impact on the behavior of public employees. "Like most city government, ours did not think in terms of business units or costs. We used standard government accounting principles that prevented our managers from stealing money, but we did nothing to stop them from wasting it. We tracked the amount of money spent on salaries, equipment, capital investments, and professional service contracts, but did not break down any of those costs by individual activities. As a direct result, city employees neither knew nor cared about their costs of doing business." After implementation of ABC, employees have been more willing and able to make cost-effective decisions, Goldsmith argues.

E. DFAS's Recent Experiences with the A-76 Cost Comparison Process

In this appendix, we discuss DFAS's recent experiences with the A-76 cost comparison process. This process is argued to be a worthwhile direction for government agencies like DFAS to reduce their fixed and incremental costs. "A-76" refers to the Office of Management and Budget (OMB) circular governing most sourcing cost comparisons between provision by government-employed civilians and by contractor personnel.

In the mid-1990s, DFAS started an A-76 cost comparison involving Out of Service (OOS) Debt operations, a collection agency–type function whereby DFAS pursues individuals who have left the armed services owing money to the government. This A-76 cost comparison was ultimately canceled, but the proposed government employee Most Efficient Organization (MEO) was implemented. As a result, in 1997, DFAS consolidated most of its OOS Debt operations in Denver. To the extent that fixed costs are location-driven (i.e., fixed costs are incurred whenever an output is performed at a different location), consolidation might be expected to reduce overall fixed costs.

As measured from Denver's perspective, this consolidation has been a considerable success. Figure E.1 shows Denver OOS Debt cost and workload data. Concurrent with the consolidation of workload in Denver, OOS Debt workload (the broken line) there went up by nearly a factor of 10 but Denver OOS Debt costs (the solid line) have only marginally increased. Obviously, Denver personnel changed their OOS Debt process in a major way to accommodate this workload increase. Such reform is a hoped-for effect of A-76 competition.

Note that this OOS Debt result is consistent with our earlier assertion about the apparently fixed nature of DFAS's costs more generally. Costs do not increase in proportion to workload increases.

Meanwhile, as OOS Debt work largely consolidated in Denver, Cleveland, Indianapolis, and Kansas City lost their OOS Debt workload—and ultimately eliminated (or at least reallocated) the associated costs. Cleveland stopped OOS Debt expenditures in early 1997, Indianapolis at the end of 1996, and Kansas City in mid-1997. Figure E.2 shows that combined OOS Debt expenditures for these three regions plus Denver fell markedly in 1997 when the consolidation occurred.

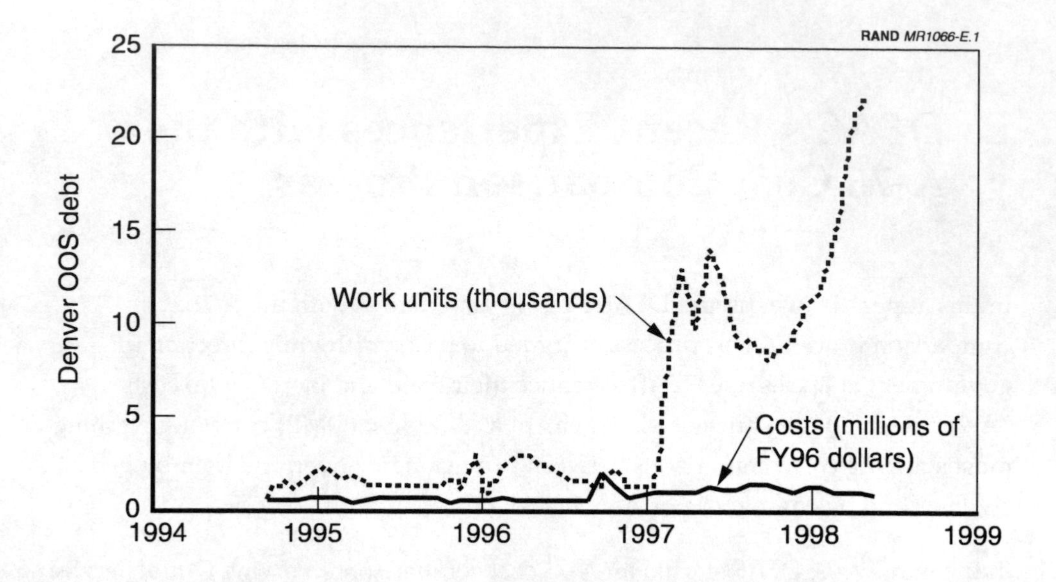

Figure E.1—Denver OOS Debt Costs and Work Units

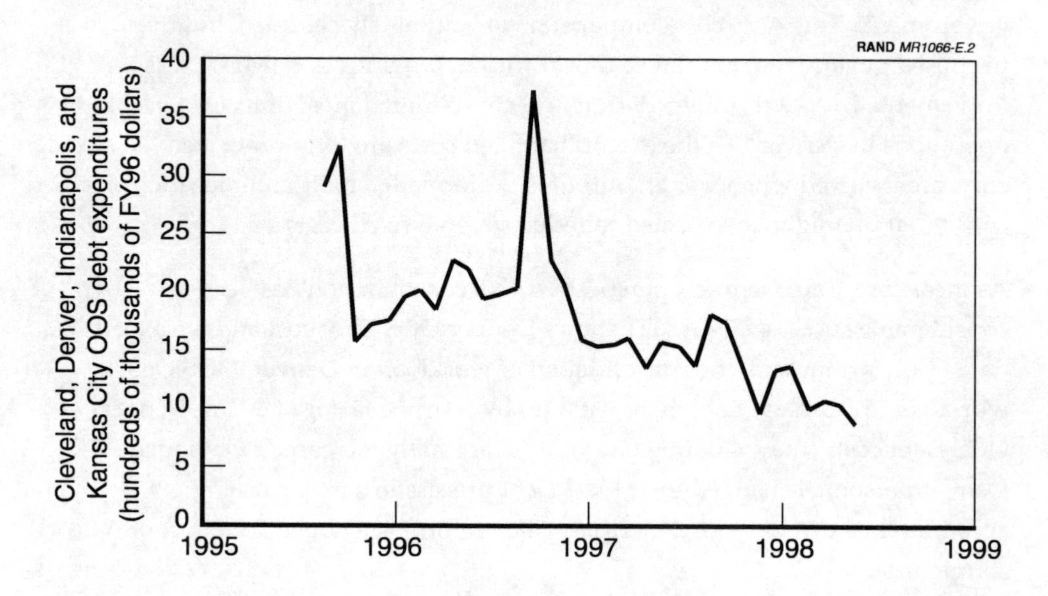

Figure E.2—Cleveland, Denver, Indianapolis, and Kansas City OOS Debt Costs

Because RADSS is a cost and work unit database, it does not supply quality measures needed to fully assess whether OOS Debt consolidation did, in fact, yield efficiency gains. However, at least from a cost perspective, this A-76-driven consolidation appears to have been a major success.

One caution, however, is that OOS Debt was a small output in Cleveland, Indianapolis, and Kansas City before its ultimate removal. Hence, it might not have been difficult for the personnel performing that output to be transferred to other outputs without having a measurably negative impact on those outputs' performances.

For example, Table E.1 shows that OOS Debt costs didn't exceed 4 percent of total costs in Cleveland, Indianapolis, or Kansas City in November 1995 (before any OOS Debt work started to be transferred to Denver). Such relatively small costs perhaps would have been fairly easy to transfer to other outputs.

Despite the apparent success of the OOS Debt consolidation, we want to caution against unbridled optimism about the A-76 process. An example from Columbus provides a note of caution: A government-employee MEO was officially implemented in Columbus in early calendar 1998 for the Defense Commissary Agency (DeCA) vendor pay function. This function's costs are subsumed into RADSS's Contract Invoices (SAMMS) output costs. As shown in Figure E.3, Columbus' Contract Invoices (SAMMS) costs have not meaningfully decreased in the aftermath of the official DeCA vendor pay MEO implementation.[1] The location of the y axis is designed to show approximately when the MEO was officially implemented.

Table E.1

The Comparative Importance of OOS Debt

Region	November 1995 OOS Debt Costs	Total November 1995 Costs	November 1995 OOS Debt %
Cleveland	$315,904	$24,556,336	1.3
Indianapolis	$651,427	$36,227,280	1.8
Kansas City	$113,148	$3,286,729	3.4

NOTE: All costs in FY96 dollars.

[1]Figure E.3 covers August 1995 through September 1998. We extended our RADSS data analysis for this figure to see if costs have decreased since the official MEO implementation date; they have not.

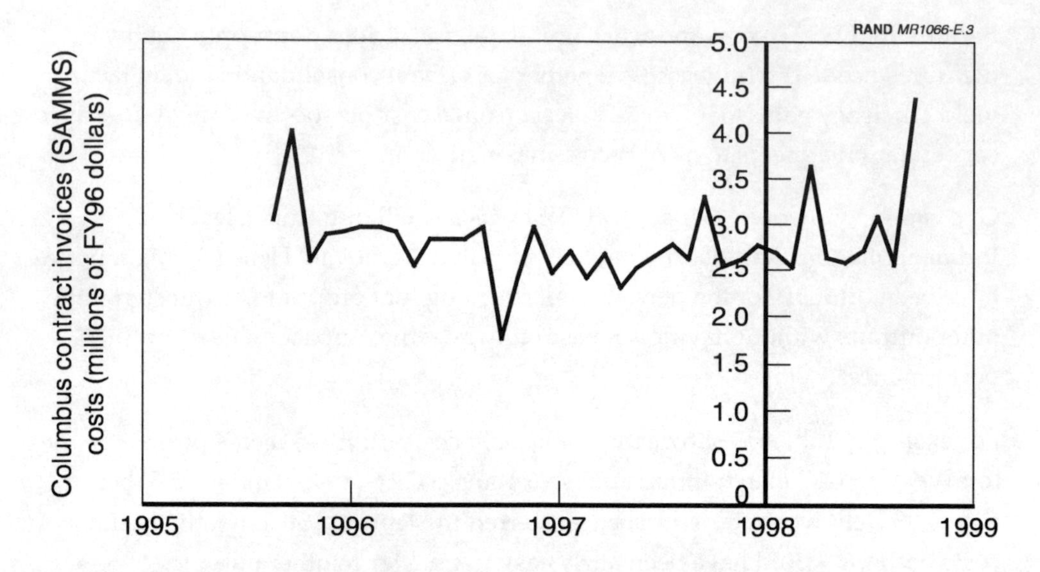

Figure E.3—Columbus Contract Invoices (SAMMS) Costs

Columbus resource management personnel told us the explanation for this phenomenon is that the government-employee DeCA vendor pay MEO was, in fact, implemented well in front of the official implementation date. Hence, one does not see dramatic cost reductions associated with the official MEO implementation.

This anecdote suggests one cannot necessarily expect to see large-scale cost savings emanating from A-76 studies at the time of study outcome implementation—particularly, one suspects, if incumbent government employees win the cost comparison. Obviously, this type of outcome considerably complicates measurement of A-76–driven cost savings.

We are currently awaiting evidence from other DFAS A-76 studies as to whether they have been as successful as the OOS Debt effort appears to have been.

References

Abraham, Katharine G., John S. Greenlees, and Brent R. Moulton, "Working to Improve the Consumer Price Index," *Journal of Economic Perspectives*, Vol. 12, No. 1, Winter 1998, pp. 27–36.

Baldwin, Laura H., and Glenn A. Gotz, *Transfer Pricing for Air Force Depot-Level Reparables*, Santa Monica, CA: RAND, MR-808-AF, 1998.

Cokins, Gary, Alan Stratton, and Jack Helbling, *An ABC Manager's Primer*, New York: McGraw-Hill, 1992.

Cooper, Robin, and Robert S. Kaplan, "Measure Costs Right: Make the Right Decision," *Harvard Business Review*, September–October 1988, pp. 96–103.

Cooper, Robin, and Robert S. Kaplan, "Profit Priorities from Activity-Based Costing," *Harvard Business Review*, May–June 1991, pp. 130–135.

Cooper, Robin, and Robert S. Kaplan, "The Promise—and Peril—of Integrated Cost Systems," *Harvard Business Review*, July–August 1998a, pp. 109–119.

Defense Finance and Accounting Service Web Site: http://www.dfas.mil/index.htm.

Defense Finance and Accounting Service, *1997 Customer Service Plan*.

Defense Finance and Accounting Service, *Cost Account Code Dictionary*, June 15, 1997.

Defense Finance and Accounting Service, Deputy Director for Resource Management, *Administration of Unit Cost*, DFAS 7045.17-R, October 1996.

Defense Finance and Accounting Service, *The Administration of Unit Cost, FY 1998 Work Count Guidance*.

Department of Commerce, Bureau of Economic Analysis Web Site: http://www.bea.doc.gov.

Department of Defense, *Annual Report to the President and the Congress*, 1998.

Department of Defense, Office of the Under Secretary of Defense (Comptroller), *A Plan to Improve the Management and Performance of the Department of Defense Working Capital Funds*, September 1997.

Department of Defense, *Report of the Commission on Roles and Missions of the Armed Forces*, May 24, 1995.

Department of Defense, *Report of the Defense Science Board 1996 Summer Study on Achieving an Innovative Support Structure for 21st Century Military Superiority: Higher Performance at Lower Costs*, November 1996.

Department of Defense, *Report of the Quadrennial Defense Review*, May 1997.

Department of Defense, *Report on the Bottom-Up Review*, 1993.

Eccles, Robert G., *The Transfer Pricing Problem: A Theory for Practice*, Lexington, MA: Lexington Books, 1985.

Eccles, Robert G., and Harrison C. White, "Price and Authority in Inter-Profit Center Transactions," *American Journal of Sociology*, Vol. 94, Supplement, 1988, pp. S17–S51.

General Accounting Office, *Defense Business Operating Fund: Improved Practices Are Needed to Set Accurate Prices*, Washington, DC: GAO/AIMD-94-132, June 1994.

General Accounting Office, *Defense Infrastructure: Budget Estimates for 1996-2001 Offer Little Savings for Modernization*, Washington, DC: GAO/NSIAD-96-131, April 1996.

General Accounting Office, *Defense Infrastructure: DoD's Planned Finance and Accounting Structure Is Larger and More Costly Than Necessary*, Washington, DC: GAO/NSIAD-95-127, September 1995.

General Accounting Office, *Defense Outsourcing: Better Data Needed to Support Overhead Rates for A-76 Studies*, Washington, DC: GAO/NSIAD-98-62, February 1998.

General Accounting Office, *DoD Information Services: Improved Pricing and Financial Management Practices Needed for Business Area*, Washington, DC: GAO/AIMD-98-182, September 1998.

General Accounting Office, *DoD Infrastructure: DoD Is Opening Unneeded Finance and Accounting Offices*, Washington, DC: GAO/NSIAD-96-113, April 1996.

General Accounting Office, *Financial Management: DoD Inventory of Financial Management Systems Is Incomplete*, Washington, DC: GAO/AIMD-97-29, January 1997.

General Accounting Office, *Foreign Military Sales: DoD's Stabilized Rate Can Recover Full Cost*, Washington, DC: GAO/AIMD-97-134, September 1997.

General Accounting Office, *High Risk Series: Defense Financial Management*, Washington, DC: GAO/HR-97-3, February 1997.

General Accounting Office, *Privatization: Lessons Learned by State and Local Governments*, Washington, DC: GAO/GGD-97-48, March 1997.

General Accounting Office, *Travel Process Reengineering: DoD Faces Challenges in Using Industry Practices to Reduce Costs*, Washington, DC: GAO/AIMD/NSIAD-95-90, March 1995.

General Accounting Office, *Financial Management: An Overview of Finance and Accounting Activities in DoD*, Washington, DC: GAO/AIMD-97-61, February 1997.

Goldsmith, Steven, "Can Business Really Do Business with Government: The Answer Is Yes, Just Ask the Mayor of Indianapolis," *Harvard Business Review*, May–June, 1997, pp. 110–121.

Harris, M., C. H. Kriebel, and A. Raviv, "Asymmetric Information, Incentives and Intrafirm Resource Allocation," *Management Science*, Vol. 28, No. 6, June 1982, pp. 604–620.

Hausman, J., and J. Mackie-Mason, "Price Discrimination and Patent Policy," *RAND Journal of Economics*, Vol. 19, No. 2, Summer 1988, pp. 253–265.

Hirshleifer, J., "On the Economics of Transfer Pricing," *The Journal of Business*, Vol. 29, 1956, pp. 172–184.

Johnston, J., *Econometric Methods*, New York: McGraw-Hill Book Company, 1984.

Kaplan, Robert S., and Robin Cooper, *Cost and Effect: Using Integrated Cost Systems to Drive Profitability and Performance*, Boston, MA: Harvard Business School Press, 1998b.

Katz, M., "Nonuniform Pricing, Output and Welfare under Monopoly," *Review of Economic Studies*, Vol. 50, 1983, pp. 37–56.

Kennedy, Peter, *A Guide to Econometrics*, Cambridge, MA: The MIT Press, 1993.

Malinvaud, Edmond, "Decentralized Procedures for Planning," in M.O.L. Bacharach and E. Malinvaud, eds., *Activity Analysis in the Theory of Growth and Planning: Proceedings on a Conference Held by the International Economic Association*, New York: St. Martin's Press, 1967, pp. 170–208.

Moulton, Brent R., "Bias in the Consumer Price Index: What Is the Evidence?" *Journal of Economic Perspectives*, Vol. 10, No. 4, Fall 1996, pp. 159–177.

National Performance Review, *From Red Tape to Results: Creating a Government That Works Better and Costs Less*, Washington, DC: Government Printing Office, 1993.

Ng, Yew-Kwang, and Mendel Weisser, "Optimal Pricing with a Budget Constraint—The Case of the Two-Part Tariff," *Review of Economic Studies*, July 1974, pp. 337–345.

Oi, Walter, "A Disneyland Dilemma: Two-Part Tariffs for a Mickey Mouse Monopoly," *Quarterly Journal of Economics*, Vol. 85, No. 1, February 1971, pp. 77–96.

Robbert, Albert A., Susan M. Gates, and Marc N. Elliott, *Outsourcing of DoD Commercial Activities: Impacts on Civil Service Employees*, Santa Monica, CA: RAND, MR-866-OSD, 1997.

Solomons, David, *Divisional Performance: Measurement and Control*, Princeton, NJ: Markus Wiener Publishing, Inc., 1965.

Spence, A., "Nonlinear Prices and Welfare," *Journal of Public Economics*, Vol. 8, 1977, pp. 1–18.

Varian, H., "Price Discrimination and Social Welfare," *American Economic Review*, Vol. 75, 1985, pp. 870–875.

Varian, H., "Price Discrimination," in *Handbook of Industrial Organization*, R. Schmalansee and R. Willig, eds., Amsterdam: North-Holland, 1987, pp. 597–654.

Weitzman, Martin, "Price Versus Quantities," *Review of Economic Studies*, Vol. 41, 1974, pp. 477–491.

Williamson, Oliver, "Hierarchical Control and Optimal Firm Size," *Journal of Political Economy*, Vol. 75, No. 2, 1967, pp. 123–138.